BUSINESS CALCULUS TODAY:
with TI-85 Graphics Calculators®

PRELIMINARY EDITION

Robert L. Richardson
Dona F. Alejandro

Appalachian State University

Saunders College Publishing
HARCOURT BRACE COLLEGE PUBLISHERS

Fort Worth   Philadelphia   San Diego   New York   Orlando   Austin
San Antonio   Toronto   Montreal   London   Sydney   Tokyo

Copyright ©1996 Harcourt Brace & Company

All rights reserved. No part of this publication may be reproduced or transmitted in any form or by any means, electronic or mechanical, including photocopy, recording, or any information storage and retrieval system, without permission in writing from the publisher.

Requests for permission to make copies of any part of the work should be mailed, to: Permissions Department, Harcourt Brace & Company, 6277 Sea Harbor Drive, Orlando, Florida 32887-6777.

TI-85 Graphics Calculator and TI-82 Graphics Calculator are trademarks of Texas Instruments, Inc.

Cover credits: Graphics calculator photo courtesy of Texas Instruments; top left: Andy Sacks/Tony Stone Images; middle: Kaluzny-Thatcher/Tony Stone Images; bottom right: Charles Thatcher/Tony Stone Images.

Printed in the United States of America.

Richardson & Alejandro; <u>Business Calculus Today: With TI-85 Graphics Calculator, Preliminary Edition.</u>

ISBN 0-03-017558-5

567 017 987654321

TABLE OF CONTENTS

Preface
Notes to Student

CHAPTER 1: DERIVING FUNCTIONS FROM DISCRETE DATA SETS
1.1 Discrete Data 1
1.2 Fitting Curves To Data 8
1.3 Finding The Curve Of Best Fit 15
Construction Industry Problem 20

CHAPTER 2: INTRODUCTION TO BUSINESS TERMS
2.1 Review Of Linear Equations 23
    Point-slope form 25
    Slope-intercept form 25
2.2 Demand, Supply, And Market Equilibrium 32
2.3 Revenue, Cost, Profit And Break-Even Point 41
2.4 Roots And Solving Equations 48
2.5 Summary And Tracing To Get Maximums And Minimums 59

CHAPTER 3: DIFFERENTIAL CALCULUS
3.1 Introduction To Marginal Concept 68
3.2 Graphing And Further Marginal Considerations 79
    Shrinking and expanding the data range 79
    Scaling– one set of values is negative, the other positive 85
    Average cost 89
The Bicycle Problem 92
3.3 Introduction To The Derivative 94
3.4 The Derivative 102
3.5 Finding Some Formulas And Theorems 110
    Elementary derivative number 1: 110
    Theorem 1: (constant times a function) 111
    Theorem 2: (sum and difference) 111
3.6 Product And Quotient Theorems 114
    Theorem 3 (product theorem): 114
    Theorem 4 (quotient theorem): 115
3.7 The Chain Rule 118
    Theorem 5 (chain rule theorem): 118
3.8 Using The Derivative To Find Marginal Values And Peaks And Valleys Of A Function 123
3.9 Maximums And Minimums With Constraints 134
    Maximum profit with constraints 135
    More on maximum profit with constraints 138
    Economic order quantity, batteries 139

CHAPTER 4: EXPONENTIAL AND LOGARITHMIC FUNCTIONS
4.1 Future Values, Present Values, And Annuities 147
4.2 Annuities: Future And Present Values 155
4.3 Amortization, Calculating Payments 164
4.4 Additional Differentiation (ln x and e^x) 169
4.5 When The Derivative Fails: Using Limits To Find Peaks And Valleys 178

i

    Gompertz Model 178
    The modified exponential growth curve 179
    Logistic models 180
    Normal curves 181

CHAPTER 5: APPLYING CALCULUS CONCEPTS TO DISCRETE DATA
  5.1 Definition Of Continuity 185
  5.2 Graphing Piece-Wise Functions 192
       Jump discontinuity 192
       Economic order quantity (continuous) 197
       EOQ (discontinuous) 199
       Using the "int" function 201
  Bookstore Inventory Planning   208
  5.3 Elasticity 210
  5.4 Point Elasticity Of Demand 231
  5.5 Further Graph Analysis: The interaction between a function and its first
       and second derivatives 237
       Point of diminishing returns 241
  5.6 Using Calculus Concepts On Discrete Data 244

CHAPTER 6: INTEGRAL CALCULUS
  6.1 Anitdifferentiation 257
  6.2 Area And The Fundamental Theorem Of Calculus 266
       Oil Problem 266
       Fundamental Theorem Of Calculus 272
  6.3 Probability Density Functions 277
       Distribution 277
       Standard Deviation 278
       Probability for independent events: 281

CALCULATOR INSTRUCTIONS .... **Appendix A**
ANSWERS TO ODD NUMBERED PROBLEMS .... **Appendix B**

Preface

This text refashions and revitalizes the business calculus course. Rather than presenting a watered down engineering calculus course to business students, we present topics proven useful in business. A draft has been used by many students at Appalachian State University over the past few years with great success.

Numerous examples and exercises include real-world applications with realistic, often discrete, data. Conceptual understanding is emphasized over symbolic manipulation. Technology is integrated throughout this text and students learn how and when to take advantage of these tools. Using this technology gives students the tools for computation and frees them to concentrate on analysis, interpretation, and synthesis.

The instructions in this version of the text are designed with the capabilities of the TI-85 graphics calculator in mind. (A version of this text for spreadsheets and Derive is also available.) The TI-85 Link was used extensively in this text and the answer keys; the graphs appear as you see them on the screen of the TI-85. Although TI-85 references and instructions are made throughout the text, major operation instructions needed for this course are located in Appendix A. The TI-82 graphics calculator can be used with this text, but in some situations students using the TI-82 are at a disadvantage. Instructions and suggestions for using the TI-82 with this text are given in Appendix A.

Answers to selected exercises are given at the back of the book. The Instructor's Resource Manual includes solutions to all exercises as well as a sample schedule, and comments and suggestions on: use of TI-85 and TI-82 graphics calculators, testing, homework, display units, TI-Link and downloading programs from Texas Instruments' archive on the Internet, and specific chapters and sections.

We hope you and your students enjoy using this text. If you have comments or suggestions, or if you find errata, please let us know at the address below. We look forward to hearing from you.

Robert L. Richardson July 1995
Dona Flake Alejandro

Department of Mathematics
Appalachian State University
Boone, NC 28608
e-mail: richardsonrl@conrad.appstate.edu

Notes to Student

Using this text will be a new and different experience for most of you. Your training to date has stressed manipulation, usually at the expense of dealing with problems of any significance. However appropriate this approach is at lower levels of mathematics instruction, you should reach a point (**before** you finish your last math course) when you spend time thinking about, and setting up, significant problems.

This text will encourage you to spend that time by giving you interesting problems to solve and by integrating technology into your solution methods. You will quickly find that the emphasis that had been placed (rightly or wrongly) on manipulating symbols is now shifted to "quiet time" thinking. For the most part, manipulation will be relegated to the technology available. You will be the supervisor, deciding what needs to be done next and how to do it. It may take you hours to plan your strategy, then 1-3 seconds to execute it by machine. This approach emphasizes **thinking and understanding.**

Many of the examples and exercises in this text use realistic data (often taken from business journals and newspapers). These real-world problems model a wide variety of business and economics topics: profit and loss, elasticity, revenue and demand functions, equilibrium prices, break-even point, marginal concept, compound interest, taxation, unemployment rates, business cycles and indexes, inventory planning, etc. (Some of these topics may already be familiar to you from other courses.)

Some of the problems included here have no absolutely "correct" solution, only a solution that best suits your reasoning. Your ability to reason will become better and better as the course progresses. Writing justifications or interpretations of your answers will be an integral part of much of your work. You should find that the ability to explain your answers and reasoning is crucial in later courses and in your professional life. Based on students' responses to this manuscript at our school, we think you will find that solving significant, realistic problems using technology can also be fun!

Robert L. Richardson
Dona Flake Alejandro

July 1995

Note: When you see the calculator symbol:

 Calculator Instructions I - Entering data points in the STAT editor

you should refer to Appendix A or to handouts from your instructor for your specific calculator. In the example above, the instructions would tell you how to enter discrete data in the STAT editor of your calculator.

CHAPTER 1: DERIVING FUNCTIONS FROM DISCRETE DATA SETS

Section 1.1 Discrete Data

Until now most of you have dealt with the world of mathematics as viewed by mathematicians and, in particular, "text book" mathematics. In text book mathematics, the world is full of problems with nice answers (note that I have implied here that not only is every answer nice, every problem HAS an answer). Problems that occur have relatively simple continuous functions to work with. I have chosen as part of my task to bring you into the world of reality, a world in which most problems have several different answers or NO answer at all; if you do get an answer you cannot be absolutely positive it is correct, and it will rarely be NICE!

You are all familiar with the arithmetic of definite answers: eg., 2+3=5. In addition, you have learned to do some fairly neat tricks with algebra (I hope): solve $3x-1=2$; factor x^2-3x-4, etc. But imagine, if you can, your accountant telling you that profit followed the curve $5x-200$. If you **cannot imagine** that happening you are probably ready to use this book!

You should be aware that the real world deals with discrete data collections much more often than with "nice" functions made up by a math teacher. What is discrete data? It means a collection of points (often ordered pairs) in the general sense. For example, suppose sales in thousands of units for the last 5 months are as follows:

Example 1: (A sales problem)

Month	January	February	March	April	May
Sales volume	2.3	2.2	2.15	2.13	2.12

What would be a reasonable prediction for June? Your set of given points would be:

(January, 2.3)
(February, 2.2)
(March, 2.15)
(April, 2.13)
(May, 2.12)
You wish to have some basis for predicting the pair (June, ?).

Example 2: (A pricing problem)

As another example, consider the following table showing the relationship between the price p in dollars per unit and the volume of sales x thousand units when a company introduced its new product in seven comparable cities. The data was compiled over a period of three months. What would be a reasonable prediction for some comparable CITY 8 if the product is to be introduced at a price $p=10.5$?

	CITY 1	CITY 2	CITY 3	CITY 4	CITY 5	CITY 6	CITY 7
x	19.5	20.5	21.1	22.9	23.2	24	25.5
p	13	12.25	12	11.5	11.25	11	10.75

Now you have seven data items and your set of points is:

(21.1,12), (19.5,13), (22.9,11.5), (25.5,10.75), (20.5,12.25), (23.2,11.25), (24,11).

Notice the following:

1. That our data is discrete, i.e., a collection of points.
2. There is no uniquely "correct" answer to either problem in the sense that you should bet your life on whatever answer you get.
3. External things (such as the economy, advertising effectiveness, availability, etc.) may very well change or even invalidate any solution we do get.

Given all these variables, what is the point of using mathematics to try to resolve these problems? The answer is that by using mathematics correctly we can get a "best" answer in some sense under the given conditions.

Nearly all of us have had to solve problems like this, usually in some kind of hit-or-miss fashion. Most of you have never had to solve a problem like this using mathematics. How does one approach these problems and find solutions?

That is what our first chapter is about. We will be using many tools of mathematics that you have never heard of. We will be doing computational things on a calculator in seconds that you could not possibly do right now in years. In short, we will be beating these problems to death with every tool at our disposal. You may think you know some math. Would it surprise you to know that nearly all the math you know and more was published in a book titled "business arithmetic" in the year 999? Unless you have had some unusual teachers, all of the mathematics you know is a thousand years old, much of it going back to the Greeks 2,000 years ago!

Consider again the sales data given in example 1.

Month	January	February	March	April	May
Sales volume	2.3	2.2	2.15	2.13	2.12

To analyze a collection of data such as this, a first step is to graph it. This is relatively simple, but since we will be doing much analysis on some fairly large sets of data, we might as well learn how to enter and graph this data in a graphics calculator.

At this juncture I would like you to note the following. Whereas I will display the results for you in the text, and your instructor will show you how to do the work using a calculator, you should work each and every example yourself to make sure you can arrive at the same results with your calculators. Getting a firm grip on the mathematical concepts involved and learning effective use of your calculator is crucial. As your instructor does a good job of the demonstration you may be led into a false sense of security. Your instructor may make it look so easy that you do not practice yourself. You come to rely on what you have seen and the calculator. **MISTAKE! MISTAKE!** Doing that is like watching a tennis instruction program and because it looks so easy you think you can do it right off. The instructor can show you exactly how to do it, but you will not be competent until you have <u>done it yourself</u> **and practiced** several times.

Calculator Instructions I - Entering data points in the STAT editor and displaying data points graphically

In this first example we want to get an "xyLine" graph. The term "xyLine" graph means you will plot the given data as coordinate pairs and join each pair with a straight line segment. Our pairs are (January, 2.3), February, 2.2) etc., but since we wish to use an XY graph, numbers are required. Hence, we will replace the months with numbers. January will be assigned 1, February 2, etc., yielding the ordered pairs:

(1, 2.3), (2, 2.2), (3, 2.15), (4, 2.13), (5, 2.12).

Now enter this data in the STAT editor of your calculator, graph a scatter plot, and then an xyLine graph. As noted in handout I, before graphing (DRAW) in the STAT editor you must clear or deselect any functions in the GRAPH editor. You must also remember to set the RANGE to include your data points. For a first look at this data, set the RANGE as follows:

xMin=0   yMin=0
xMax=6   yMax=3
xScl=1   yScl=1

The graphs that you get may look like these:

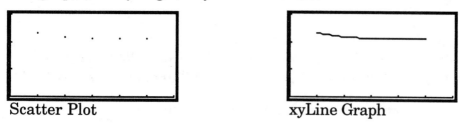

Scatter Plot   xyLine Graph

Since the Y-axis numbers are relatively close together, the graph looks nearly like a straight line when you vary the Y-axis from 0 to 3. If you varied the Y-axis from 0 to 10, your graph would look even more like a straight line. Not a very helpful picture, is it? Now regraph using a Y-Min of 2, a Y-Max of 2.6, and a scale of .1. The graphs that you get may look like these:

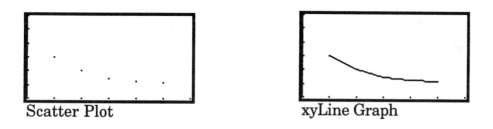

Scatter Plot   xyLine Graph

Note: the dots along the bottom of the graph indicate that the graph window is above the x-axis. When turning in work, your graphs should be labeled and the scale should be indicated. I have illustrated this on the next two graphs in this section.

Another relatively easy way to graph the data in example 1 is to subtract a common value from each entry, permitting you to change the scale. Looking at the sales volume numbers which yield the Y-axis values, we see that if "2" is subtracted from each entry the resulting numbers vary only from 0 to about .3, yielding a more noticeable change in the graph values. The data points entered into the STAT editor should be: (1, .3), (2, .2), (3, .15), (4, .13), (5, .12). The range window can then be set as: xMin=0, xMax=5, xScl=1, yMin=0, ymax=.3, yScl=.05. Name these lists as xSTATE1 and ySTATE1; you will use these again. The graphs that you get may look like these:

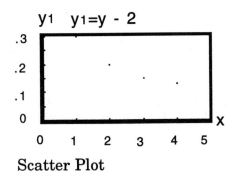
Scatter Plot

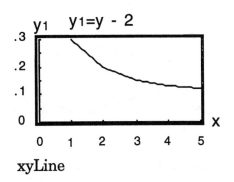
xyLine

You may wonder if something has been lost in doing this. While it is true that you could mislead a client with such a graph; emphasizing differences, when done carefully, will prove invaluable to us in analyzing what is happening because we are really interested in the difference between points. We look for trends and change.

Calculator Instructions I - Entering and plotting two sets of data on the same screen

As another example, consider the situation which arises when you have two sets of data to be plotted against each other. Suppose we consider the data we have for certain 5-year periods and some 1-year periods to see if there might be any connection between the number of moves made by a moving company and a certain consumer price index. It is almost always the case that some graphical display of the information plotting the two results to be considered against each other on a vertical axis will give some insight into the situation.

Example 3:

X-axis	1975	1980	1985	1990	1991	1992	1993	1994
Y1-axis	226	257	221	250	248	247	240	251
Y2-axis		1.2	1.8	2.01	2.1	1.95	2.3	1.9

Since most calculators do not have a double-ended y-axis capability, you will need to do some scale adjusting before entering this data. If you were to subtract 220 from each of the Y1 values and 1975 from each of the X values, your resulting data sets would be as follows:

Moves - {(0, 6), (5, 37), (10,1), (15,30), (16, 28), (17, 27), (18, 20), (19, 31)}
Index - {(5, 1.2), (10, 1.8), (15, 2.01), (16, 2.1), (17, 1.95), (18, 2.3), (19, 1.9)}

Set your range window to be: xMin=0, xMax=20, xScl=1, yMin=0, yMax=38, yScl=1. Enter the data sets and draw on the same screen to get a graph that might be similar to the following:

y

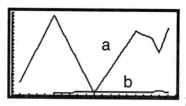

a) x=x1+1975   y1=y+220
b) x=x1+1975   y2=y

Moves vs. index

This is far from perfect but it may be the best you can do with most graphics calculators. Remember that many businesses will have computers with good spreadsheet software and/or an art department that you can go to for help to get nice 3D color graphs.

Information needed to work exercise 1.1

1. Entering discrete data in the STAT editor.
2. Displaying data points as a scatter plot and as an XY line graph.
3. Adjusting scale and Range window to get a "nice" graph.
4. Entering and plotting more than one set of y-data.

Exercise 1.1

In the following problems, enter the data, do an appropriate scale adjustment when necessary, and get a "nice" XYLine graph. Sketch the graphs on the paper; remember to label and to indicate scale and scale adjustment on the graph. In each example, use the first row (possibly adjusted) for the X-values and the second rows (possibly adjusted) for the Y-values. Finally, **NAME YOUR DATA SETS** as you will be doing further work on these problems. For example: for number one, name as xSTATP1 and ySTATP1 in the STAT editor.

1. Graph the data set in Example 2 (A pricing problem). The volume of sales x (in thousand units), with a possible adjustment, should be used for your x-axis and the price p, with a possible adjustment, should be used for your y-axis.

2. Create "nice" XYLine graphs by scale adjusting one or both of the Y data. Assign the values 5 to May, 6 to June, 7 to July, etc. Use these values (5,6,7,8,9) for your X-axis.

X-axis	May	June	July	August	Sept.
Y1-axis	1993.8	1993.5	1995.4	1997	1999
Y2-axis	15.1	14.2	15.8	16.3	15.7

3. In this problem, use the years (as they appear) for the X-scale. Adjust either the Y1 scale for the first row or the Y2 scale for the second row.

X-Scale	1988	1989	1990	1991	1992	1993
Y1-Scale	27.8	29.5	34.6	33.5	33.5	34.7
Y2-Scale		1.7	5.1	-0.9	0	1.2

Section 1.2 Fitting Curves To Data

We will now continue working on the data in the sales volume problem and the resulting graph. They are named as "xSTATE1" and "ySTATE1". The data points were entered as: (1, .3), (2, .2), (3, .15), (4, .13), (5, .12).

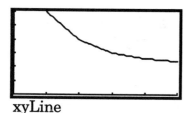
xyLine

Now that we have a good view of the function, we need to make some kind of reasonable prediction as to what will happen in the next month. There are a number of ways to do this, each involve GUESSING! ------ OOPS. Did I say guessing? What kind of math book is this anyway! You thought math was exact, right? Well, our math <u>calculations</u> WILL BE exact, but only a moment's reflection should tell you that we are essentially going to peer into a crystal ball and come up with a guess. Not just ANY guess, mind you. We want the "best" possible guess (whatever "best" means here), or, at the very least, a good guess. In fact, we may frequently try several guesses on the same problem. This is where the mathematics comes in.

Questions like this have been puzzling mankind since the beginning of mankind. We have <u>always</u> wanted to know what will happen next, hence the market for numerologists, astrologists, palmists - there may be no end to the list of "ists". Our job is to introduce the element of science into what we do as much as possible.

An element of science is interjected as soon as you clearly and carefully state what your assumptions are, then use only techniques which can be proven to be correct under those assumptions to solve the problem. We will do this in our current problem.

Assumption 1:
<u>The given points closely follow some underlying mathematical function</u>.

The word "some" here is crucial. What function? In all likelihood we will never know. That is where the "guess" part comes in. We try to consider only those functions we can deal with and which appear reasonable for the given data. In this class, we will select some of the most commonly used functions. They are shown in the following chart:

Linear: y=a+bx
b(slope)   a(y-intercept)

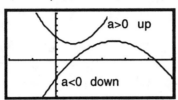

Quadratic: y=ax^2+bx+c
c (y-intercept),
-b/2a (x-coordinate of vertex)

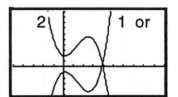

Cubic:
y=ax^3+bx^2+cx+d

4th order polynomial:
y=ax^4+bx^3+...+e

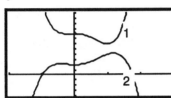

Logarithmic:
y=a+b ln x   x > 0

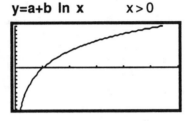

Exponential: y=ab^x
a>0 & x>0: 1(b>1) 2(0<b<1)

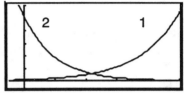

NOTE: if a>0 & x<0 then 2(b>1)
                         1(0<b<1)

Power: y=ax^b
x>0   a>0   0<b<1

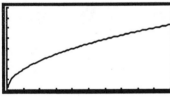

Power: y=ax^b
x>0   y>0   a>0   b<0

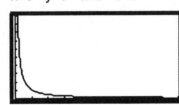

Power: y=ax^b
x>0   a>0   b>1

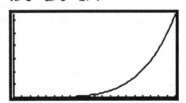

The last three are not choices we can use for regression analyses but they are graphs that we will use in later chapters.

y=ae^[-b(x-c)^2]

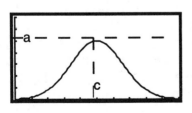

y=a(1-e^(-bx))

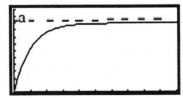

y = (bx+a)/x

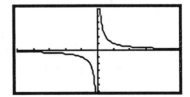

You should already be familiar with most of these functions, but perhaps not their graphs. If you have been through a traditional math program you will need to be particularly observant of this chart (memorize it?) and draw appropriate generalizations about the relationships between functions and their graphs as you work. You would have seen and graphed a straight line, y=ax+b so we won't discuss that one.

How about the quadratic (a polynomial of degree 2)? Note that a quadratic can open up or down. This is determined by the coefficient of x^2 - if it is negative the curve opens down (often called "concave down") and if the coefficient is positive the curve opens up ("concave up"). An interesting general characteristic of polynomials is that in general a polynomial of degree n has n-1 bends or concavities. Under special circumstances these can collapse to fewer. For example, $y=x^4$ has only one bend or concavity and looks nearly identical to $y=x^2$. You will easily see why there is this n to n-1 relationship in a polynomial as soon as you learn a little calculus.

Looking at the graph we have and our possible choices, the simplest one I see that looks like the given data is a quadratic. I will choose that.

Assumption 2:
  The given points closely follow a quadratic equation.

There may well be extraneous circumstances not mentioned in our problem that would cause me to choose another, such as a cubic. As you will soon see, the choice of a function as simple as a quadratic will be all we will want to handle for now. In fact, without computerized help we would be unable to make that choice in this class as the concepts are not easy and the computations are an absolute nightmare.

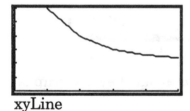
xyLine

Looking at the graph of the data we observe several possible choices for functions. Looking at the functions in our graphing chart, y=(bx+a)/x looks like a good choice (our best?) as well as ae^{bx}, b<0. Another not so obvious choice might be a part of a quadratic, if you chop off just the right piece. This immediately brings us to our current limitations. It will be easier and a little more illustrative for us to look at the quadratic even though that might not be the ideal choice. Also, given our limited knowledge of the situation (we didn't tell you where this data came from) one choice may be as reasonable as the other. Hence, we will proceed with the quadratic choice.

We now wish to fit a quadratic to this data. But how? It is time for you to review some techniques you have learned (but probably not applied) and to learn a few new tricks. For the moment, let us suppose we have found the required quadratic

$y=ax^2+bx+c$, where we selected the month values 1,2,3,4, and 5 for the x values. What property do you know of that the quadratic MUST have?

1	2	3	4	5
2.3	2.2	2.15	2.13	2.12

It MUST satisfy the following set of conditions:

$2.3 = a(1)^2+b(1)+c$
$2.2 = a(2)^2+b(2)+c$
$2.15 = a(3)^2+b(3)+c$
$2.13 = a(4)^2+b(4)+c$
$2.12 = a(5)^2+b(5)+c$

Why? Because if this quadratic function EXACTLY fits our curve, each time you put in an X-value you MUST get the corresponding Y-value. You can see that our problem has "simplified" to one of finding appropriate values for a,b, and c.

At first glance this does not appear to be difficult. If you expand out the equations and rewrite them as:

$2.3 = a + b + c$
$2.2 = 4a + 2b + c$
$2.15 = 9a + 3b + c$
$2.13 = 16a + 4b + c$
$2.12 = 25a + 5b + c$

you see that you simply have five linear equations in three unknowns: a, b, and c. In fact, all of you should know how to approach this problem.

To refresh your memory: when you have a collection of linear equations, you can solve them by elimination. For those of you who are a little rusty let's run thorough a quick outline of how this is done. First you can eliminate the "a" from the first two equations (or "b" or "c" if you so wish):

$2.3 = a + b + c$
$2.2 = 4a + 2b + c$

multiply the first equation by -4: $-9.2 = -4a - 4b - 4c$ and add to the second equation:

$-9.2 = -4a - 4b - 4c$
$2.2 = 4a + 2b + c$

$-7 = 0 - 2b - 3c$

This yields a new equation with no "a" in it. Continuing in this fashion you would now choose the first and third and eliminate the "a", yielding two new equations involving only "b" and "c". Using these two equations and eliminating one of "b" or "c" you would get one equation in one unknown. Working your way back up the system you would get the other two coefficients. In fact, nice fellow that I am, I have done this for you, yielding:

a=1/40, b=-7/40, c=49/20.

So we have found our quadratic equation: $y = f(x) = \frac{1}{40}x^2 - \frac{7}{40}x + \frac{49}{20}$. Hurray! But wait, what about the last two ordered pairs? Those are where the X-values are 4 and 5. We didn't need those so we didn't use them. Remember the condition this quadratic MUST satisfy. We must be able to put in any of the X values 1, 2, 3, 4, or 5 and get out the corresponding Y values. Does this happen when we put in 4 and 5? Again, I will do the arithmetic for you:

f(4) = 2.15 and f(5) = 2.2, but recall the values we originally had:

1	2	3	4	5
2.3	2.2	2.15	2.13	2.12
checks	checks	checks	2.15	2.2

This equation does not work correctly. Have we made an error? No, f(1)=2.3, f(2)=2.2, and f(3)=2.15 which is correct if we only look at the first three pairs.

What, then, is the problem? The problem is that we have TOO MUCH information to determine a unique quadratic equation. Note that y=ax² + bx + c has only 3 unknowns: a, b, and c. Thus, only the first three conditions are required to solve for these unknowns. There is no way to work in the additional information and satisfy all 5 conditions at once. How do we resolve this problem? It is time for you to learn a new concept, namely finding the **curve of best fit**. We will learn this new concept in section 1.3.

Information needed to work exercise set 1.2

1. Entering data in the STAT editor and creating an xyLine graph.
2. Viewing a graph and deciding among the curve choices you are given, which one appears to best fit your data.

Exercise 1.2

Given the following graphs or collections of points, after doing any necessary scale changes and graphing, state which type of curve you think would best fit the data. Graph each data set on your calculator and hand draw the curve on your paper; remember to label and indicate your scale. Write the general equation of the graph you chose for the "best fit". The first two problems are from exercise 1.1 and you should have saved and named them.

1. Suppose you have the following data with Y1 sales and Y2 number of employees in 1,000s. After assigning the values 5 to May, 6 to June, 7 to July, etc. Use these values (5,6,7,8,9) for your X values. Refer to problem number 2 in Exercises 1.1.

X-axis	May	June	July	August	Sept.
Y1-axis	1993.8	1993.5	1995.4	1997	1999
Y2-axis	15.1	14.2	15.8	16.3	15.7

   a. What kind of curve would best fit the [X,Y1] values?
   b. What kind of curve would best fit the [X,Y2] values?

2. In this example, use the years (as they appear) for the X-scale. Refer to problem number 3 in Exercises 1.1.

X-scale	1988	1989	1990	1991	1992	1993
Y1-scale	27.8	29.5	34.6	33.5	33.5	34.7
Y2-scale		1.7	5.1	-0.9	0	1.2

   a. What kind of curve would best fit the [X,Y1] values?
   b. What kind of curve would best fit the [X,Y2] values?

3.

1 minute	2 minutes	3 minutes	4 minutes	5 minutes	6 minutes
8	7.8	5	2	1.5	1.4

   a. What curve(s) best fit this data?
   b. Do any of the curves appear to work perfectly? Do any appear to start off well and then become really bad after a while?

4.

0 Sil.	.1 Sil.	.2 Sil.	.5 Sil.	.8 Sil.	1 Sil.
0	4	7.6	8.5	8.9	9

What curve best fits this data?

Section 1.3 Finding The Curve Of Best Fit

In section 2 we discovered that since we had too much data we could not run our first choice curve exactly through each and every point. Statisticians have a technique they use to resolve such problems. It is referred to as **the least squares method for the curve of best fit**. The idea behind this technique is not difficult to understand. In its simplest form, you try to fit a straight line to a collection of points that don't go exactly in a straight line. How can this be done and why would anyone want to do it?

An easy way to understand the reason for wanting to do this is as follows. Suppose a chemist in a laboratory makes a number of measurements at different times to see how something is changing. Due to measurement error, some or perhaps all of the observations are incorrect. The chemist, however, guesses that the observations should in fact follow a straight line. What line in some sense follows most closely the observations?

Look at the picture with the observations given as dots and the line as a guess as to what the actual line might be.

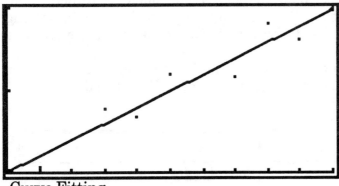
Curve Fitting

As you can see, if the data should follow a straight line, the one showed would certainly be a great candidate for the correct line. But remember I kept using the word "best" in some sense. This one looks good, but is not really best in any sense; it is just a guess. Through lengthy and careful considerations statisticians have determined that a useful choice for the unknown line ("best line") is to choose the line which has the least sum of the squared Y-distances from each observed point.

Looking at this picture we can see that the chosen line is quite good. Clearly if you were to move the line up or down much, or rotate it very much, the Y distances indicated would increase. But how exactly does one calculate the values a and b for such a line $y=ax+b$?

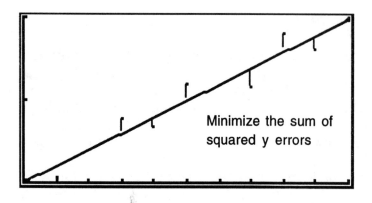

The calculations for each of the values a and b are lengthy but not difficult with the aid of a calculator. You will probably take a statistics course soon, if you have not already, and you will be required to make the necessary calculations (I STRONGLY advise you to use a calculator to do this).

The derivation of the formulas needed to make these calculations are difficult and, in fact, involve differential calculus. You will learn the general technique for solving such problems in this text. However, the derivation of the formulas is beyond the scope of our knowledge at the moment.

Now that you have the general idea of what we are going to do, let's return to our original problem. We wish to find the quadratic equation of best fit for the data.

1	2	3	4	5
2.3	2.2	2.15	2.13	2.12

To do this, we will have to find that quadratic equation which has the least squared differences between the given and calculated Y-values. Note that fitting a quadratic is much more complicated than fitting a straight line. With a line you can only slide it up and down or rotate it. You can do all of those with a quadratic equation but the equation itself can be dramatically changed. It can open up or down. It can shoot up or down steeply or in a very gradual fashion. A problem such as this is usually only solved in fairly advanced courses. In fact, it can be so complicated that I have actually seen straight lines fit to data when it was clear from both the facts and the appearance of the data that they should not have been used.

Fortunately for us, our calculator will come to our rescue.

Calculator Instructions I - Data curve fitting and forecasting

1. **Enter data points in the STAT editor.**

    Remember to set your range window. One setting might be:
    xMin=1   yMin=2
    xMax=7   yMax=2.5
    xScl=1   yScl=.1

2. **Calculate the quadratic equation of "best fit".**

    (Polynomial Regression of degree 2)
    Your calculator should return the coefficients of $y = ax^2 + bx + c$ as a=.015, b= -.1330000000003, and c=2.414.
    Therefore, our quadratic equation is $y = .015x^2 - .133x + 2.414$

3. **Draw**

    a) SCAT PLOT (Scatter Plot)
    b) REG (Regression curve)

4. **Forecast**

    We are **FINALLY** ready to make our prediction for June:

    S(6) = 2.156

1	2	3	4	5	prediction for 6
2.3	2.2	2.15	2.13	2.12	2.156

Is this correct? Who knows. Predicting future outcomes is an arcane art at best. It might seem hard to believe by just looking at the data that the predicted value for x=6 would start back up again. One of the most dangerous things you can do is to use this least squares method for drawing conclusions **beyond the boundaries of the given data**, something we just did. Statisticians are generally appalled when this is done, but it is done all the time just the same. Statisticians like to be able to predict how bad your answer could be, called the error estimate. When you use the method of least squares to predict what is going to happen beyond a certain data set you usually destroy any value the error estimate might have since it usually becomes huge. Of what value is an answer of 2.156±100? How can you minimize the possibility of a really disastrous error?

While there is no general technique that will always work, a well informed and knowledgeable user is probably the best you can do. For example, in one of the problems from the last exercise set,

0 Sil.	.1 Sil.	.2 Sil.	.5 Sil.	.8 Sil	1 Sil.
0	4	7.6	8.5	8.9	9

the type of process being considered should have an effect on our considerations for a possible curve of best fit. In this example, the kind of process being considered and how it works is crucial. This data came from a manufacturing process where a certain type of alloy was being made consisting of several different melted metals. When making an alloy of metals, one adds certain combinations of metals together to come up with characteristics within the combination better than any individual members may have. The data was gathered to see how the quantity of silver being added affected a particular characteristic of the alloy. As more and more silver was added, the characteristic being measured climbed considerably beyond what would be expected from just the silver itself. Could this kind of enhancement be expected to continue? Since most of you have never made alloys with molten metals but have played with paint, look at the following analogous situation.

What happens when you mix a teaspoon of blue paint and a teaspoon of yellow? The resulting green certainly appears different than either member. Now suppose you continue adding more and more blue. Does your mixture get more and more green? Darker? Lighter? Keep adding blue. Add 50 gallons of blue to your 1 teaspoon of yellow. Eventually what do you end up with? The answer is blue. In other words, when you are making a combination and continue adding one particular element, that element eventually overwhelms the mixing process. The same is true with alloys.

If you continue adding Si. you will end up with an alloy that has all the characteristics of Si. The other contributors have simply been overwhelmed. The correct curve for such a process would have to look like the accompanying graph where you increase the desired characteristic up to a point, and then it starts to decrease.

y-axis is the desired
characteristic

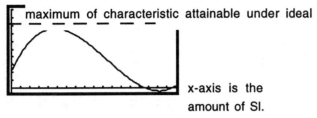

x-axis is the
amount of Sl.

Thus, if we were to fit a curve similar to this to the given points, we **MIGHT** be a little more secure in our conclusions. However, with your limited knowledge and the information I have given you, our guess of 2.156, from the forecast in step 4, would seem to be very reasonable.

Information needed to work exercise 1.3

1. On your calculator, an understanding of and ability to fit a "least squares" curve.
2. On your calculator, an ability to use the forecast feature in the STAT editor.

Exercise 1.3:

1. Find the equation of the curve of best fit for the data in number 1a, Exercises 1.2 using:

    a) a quadratic (2nd-order polynomial)
    b) a cubic (3rd-order polynomial).

2. Using the 3rd-order polynomial, in number 1 (above), what is the forecast for the sales in October?

3. Find the equation of the curve of best fit for the data in 1b, Exercises 1.2, using a 3rd-order polynomial.

4. Using the equation in number 3, what is the forecast for the number of employees for the month of October?

5. Find the least squares line for the given data. What is the forecast for y when x=25000?

x	5000	10000	15000	20000
y	50	35	27	25

6. Determine which function, linear, exponential, or power is the equation of best fit; write this equation.

a)

x	5	10	15	20	30
y	45	125	230	360	656

b)

x	2	3	5	10	15
y	8.9	13.4	30.3	230.8	1752.8

c)

x	10	20	40	50	60
y	16	36	66	82	96

7. Give the equation of the curve of best fit using a 3rd-order polynomial for problem number 3 in Exercises 1.2.

CONSTRUCTION INDUSTRY PROBLEM

The construction industry is often watched closely to see if construction is increasing or decreasing (seasonally adjusted). This is used as a leading economic indicator to see if the economy is improving or not. Given the following data, make a prediction of the spending level for the next two months. Numbers are rounded off to the nearest billion dollars. Using these numbers, if the assumption of the relation is correct, what do you predict the construction industry (and hence the economy) will do in November and December? Since this is actual data for the period, you can read ahead and see what happened. However, you should not do this since you are **predicting** what will happen based on the given data and articles in publications **prior** to November and December 1991.

Mon	1990		1991									
	N	D	J	F	M	A	M	J	J	A	S	O
Amt	435	421	408	410	401	408	399	398	400	402	408	411

Your task is to find a curve of best fit and, using it, to predict what will happen the next two months. As you know, this involves some hocus-pocus. You will **have** to research the question by using the library to find the general state of the economy at that time, the various factors that affect the construction industry, and then use extreme caution as you make your prediction. Remember as you work: this problem is from the real world and in the real world, data does not conveniently fit your mathematics, your mathematics must fit the data. Do not start patting yourself on the back about the neat math you have done and what a clever fellow you are without checking to make sure your math fits the economic climate at the time.

To get you in the right frame of mind for your report, you will place yourself in the following scenario:

Your report is to be filed with your superior. It will be read and passed along to her/his superior, ultimately (if it is not rejected as the driveling of an incompetent poop) ending up on the desk of Marvin Greenbridge, chief economist and soothsayer for the company. Marvin has arrived at his position higher than he/she is qualified for. He never really understood statistics, but is a whiz with numbers, a smooth talker, a hard worker, totally intolerant of people who do not understand statistics, and married to the CEO's daughter.

As a clever employee you are aware of all these things hence:

1. Your report will be short and to the point (Marvin is busy).
2. The explanation of your statistics work will be **VERY** clear and simple (Marvin does not really understand statistics but wants everyone to think he does).
3. Your report will contain at least one memorable statement. (Marvin likes to be known as a smooth talker and is not above quoting someone else's work, without credit of course.)
4. Your mathematics MUST be painfully accurate (Marvin is a whiz with numbers. Mistakes virtually JUMP OUT of the page at him.)
5. As a long time soothsayer, Marvin understands perfectly the guesswork involved in what he does. Your report must be sufficiently definite (Hence, we can see that ....It is my strong feeling that ...) but allow plenty of room for him to fudge (but of course, our assumptions about ... can change at any time ... it is always possible and maybe even likely that ...).
6. Marvin will expect you to have done your research **very** carefully before even beginning your mathematical analysis. He will expect references to several of the following:

The Wall Street Journal, Architectural Record, Time, Business Week, various Business Monthlies, Builder, Business Journals, government documents (frequently a special section in your library), etc. Particularly important articles should be xeroxed, important parts highlighted and included at the end of your report.

7. Marvin will expect you to consider at the very least the status of **commercial, residential,** and **governmental** construction. You might also work in other facts as you notice them: personal income, unemployment, industrial sales, etc.

8. Marvin will expect a graph of the given data. He will expect you to use some or all of this data in combination with the facts you have discovered to come up with a "best fit" curve. **This function should be stated and the predictions it gives should be overlaid on that portion of data you chose as significant in constructing the function. This function should be used to project results for November and December.** The data you choose to use from that given, and the function you choose to fit to that data, should make predictions that **agree with the narrative portion of your report** (your math must fit the real world, not the other way around.) Marvin will be extremely dismayed if your mathematics does not reflect the knowledge gained in your reading research.

9. Bad grammar (clauses and phrases for sentences, disjointed thoughts, basically all those things your English teachers have tried to teach you not to do) gives Marvin a headache.

A REVIEW OF SOME FACTS YOU LEARNED IN ENGLISH CLASS:

1. You should make an outline of what you are going to say. You may wish to deep it in the form of a diary as you work.
2. The first sentence of each paragraph should indicate what the paragraph is about.
3. The first paragraph should indicate what the report is about.
4. Such a report should include a good graph of the data.
5. And finally, FOR HEAVENS SAKE, use the spell checker before passing in your report.

CHAPTER 2: INTRODUCTION TO BUSINESS TERMS

Section 2.1 Review Of Linear Equations

There are three standard forms for the equation of a straight line. Algebraically we have what is called the "**general linear equation**" $ax+by+c=0$. "Linear" in this sense means that the equation is of degree one, which means that there are no products of variables and both x and y are in the equation with power 1. At times we will use the letter "p" for the vertical axis instead of the letter "y" and the letter "q" for the horizontal axis instead of "x".. We also need the other two forms of the equation of a straight line which are:

Point-slope Form	Slope-intercept Form
$p - p_0 = m(q - q_0)$ or $y - y_0 = m(x - x_0)$	$p = mq + b$ or $y = mx + b$

Since we will need most of this work again when we begin calculus, let's derive these three forms carefully. The basis for this derivation is an axiom from geometry which states that two points determine exactly one line. The slope of a line joining two points is defined to be the change in p over the change in q. Since p is usually the vertical scale and q the horizontal, another way to state this is that the slope of the equation is the **rise over the run**.

We first choose the pair of points (q_0, p_0) and (q_1, p_1). As indicated in figure 1.3.1, for this pair of points the rise is p_1-p_0 and the run is q_1-q_0.

Hence, the slope is: rise / run = $(p_1-p_0)/(q_1-q_0)$

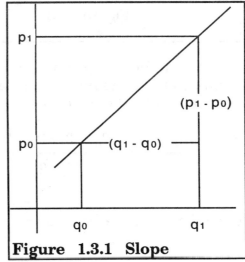

Figure 1.3.1 Slope

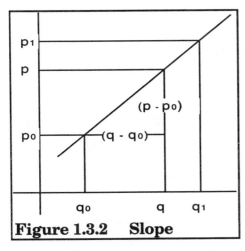

Figure 1.3.2 Slope

Using the same line but a different pair of points as indicated in figure 1.3.2, we get the rise = $p-p_0$ and the run = $q-q_0$. Hence, another way of expressing the slope for this line is: $\dfrac{\text{rise}}{\text{run}} = \dfrac{p - p_0}{q - q_0}$.

One of the features which distinguishes a straight line from all other curves is that the slope does not change as you move from place to place along the line. For example, in the curve in figure 1.3.3 below, we can consider the slope of the curve at different points by using the slopes of the straight lines tangent to the curve at the given points.

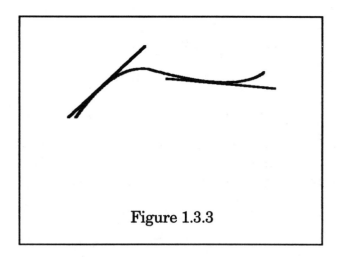

Figure 1.3.3

In our straight line in figures 1.3.1 and 1.3.2, we have merely chosen two different sets of points on the same line to calculate the slope. Therefore, the slope numbers that we get must have the same value:

$$\text{slope} = \dfrac{\text{rise}}{\text{run}} = \dfrac{p - p_0}{q - q_0} = \dfrac{p_1 - p_0}{q_1 - q_0}.$$

We can then take the fractions $\dfrac{p - p_0}{q - q_0} = \dfrac{p_1 - p_0}{q_1 - q_0}$, multiply through by $q - q_0$ and get

$$p - p_0 = \dfrac{p_1 - p_0}{q_1 - q_0}(q - q_0)$$

It is customary to use the letter "m" when referring to the slope of a straight line. Doing this and replacing $\dfrac{p_1 - p_0}{q_1 - q_0}$ by m we get the point-slope form of the equation of a straight line:

Point-slope form: $p - p_0 = m(q - q_0)$

The student just beginning calculus will not appreciate the significance of this form. Since we will be dealing with slopes for some time to come this will be an important form.

Finally, we expand and combine terms in the equation thus:

$p - p_0 = m(q - q_0)$,
$p = m(q - q_0) + p_0$,
$p = mq - mq_0 + p_0$
$p = mq + (-mq_0 + p_0)$.

Letting $-mq_0 + p_0 = b$ we get the slope-intercept form of the equation of a straight line:

Slope-intercept form: $p = mq + b$ or $y = mx + b$

Note how the formula can be interpreted. m is the slope of the line, b is the p-intercept, i.e. the point where the line crosses the p-axis (which is when q=0). This form of the equation is commonly used for graphing purposes since knowing the slope and the p-intercept will let you graph nearly instantly. This means that you should have a good "slope picture" in your mind. If you don't, memorize figure 1.3.5. Note in figure 1.3.4 how the equation: $\text{slope} = \dfrac{\text{rise}}{\text{run}}$ will be used to derive the various slopes in figure 1.3.5.

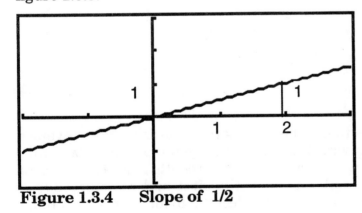

Figure 1.3.4   Slope of 1/2

$\dfrac{\text{rise}}{\text{run}} = \dfrac{1}{2} = \text{slope}$

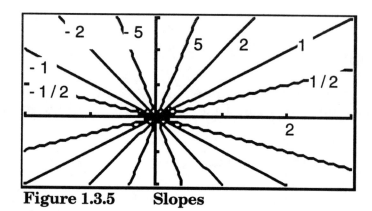

Figure 1.3.5   Slopes

Example 1: Given that a linear price - quantity relationship exists and that the price is $5 when the quantity is 100 and $6 when the quantity is 90, find the equation of the line.

Solution: We are given two pairs of points (q,p), namely (100,5) and (90,6). We identify (100,5) with (q_0,p_0) and (90,6) with (q_1,p_1) and use our formula to find the slope m.
$$m = \frac{6-5}{90-100} = \frac{1}{-10}$$
Use this slope and one of these points in the **point-slope form**:
$$p - 5 = \frac{1}{-10}(q - 100).$$
Solving for p gives us the equation in **slope-intercept form**:
$$p = -\frac{1}{10}(q - 100) + 5$$
$$p = -\frac{1}{10}q + 10 + 5$$
$$p = -\frac{1}{10}q + 15$$

A common error students make is to list the entries in the ordered pair reversed thus: (5,100). There is nothing wrong with this except that you must then swap all p and q entries in the equation or you will misinterpret the results since the equation was derived from the pair ordered (q,p) with quantity first and price second. A common question is what difference it makes if you select (90,6) for (q_0,p_0) and (100,5) for (q_1,p_1). Referring back to the way we constructed the formula from the ratios
$$\frac{p-p_0}{q-q_0} = \frac{p_1-p_0}{q_1-q_0},$$
you can see that the only result is a sign reversal in both numerator and denominator

which means that no change results.

Let us now work Example 1 by entering the two points in the STAT editor of our calculator and by finding the linear regression equation. Your screen should read as follows:

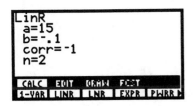 Thus we again have the equation: $p = -\frac{1}{10}q + 15$

Example 2: Given a linear price-quantity relationship with slope -2, knowing that the price is $5 when the quantity is 100, find the equation.

Solution: Using the recipe (**point-slope form**) we get: $p - 5 = -2(q - 100)$.

For **slope-intercept form**, solve to get: $p = -2q + 205$.

Significant Business Fact About Slopes:

Mathematicians are generally happy with the equation $\text{slope} = \frac{\text{rise}}{\text{run}}$ since they are concerned with general concepts. However, this equation has a very special and useful significance to business majors. You should note that $\text{slope} = \frac{\text{rise}}{\text{run}} = \frac{p_1 - p_0}{q_1 - q_0}$ actually creates a relationship between change in price (the numerator) and change in quantity (the denominator). Note that these changes are interrelated - a change in either affects the other. This "change" relationship is so important that it will be considered repeatedly in many different situations. The Greek letter Δ (delta) is often used to indicate "change in". You will frequently see $\text{slope} = \frac{\text{rise}}{\text{run}} = \frac{p_1 - p_0}{q_1 - q_0}$ expressed as $\frac{\Delta p}{\Delta q}$ when we are interested in how a change in one quantity affects or is affected by a change in the other.

© Saunders College Publishing

Calculator Instructions II - Entering and graphing functions in the GRAPH editor

Example 3: Given a linear relation with slope -110 and p-intercept 15, give the equation and graph the line.

Solution: Using the slope-intercept formula we get p=-110q+15 which gives the accompanying graph when using a standard viewing window (**ZSTD**) in the graph editor.

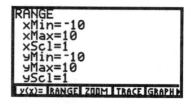
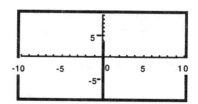

As you can see, when the slope is as nearly vertical as -110, it can be difficult to tell the difference between the vertical axis and the curve. It is therefore frequently necessary to change the scale in your graphing window to accentuate the slope of the function in question.

To adjust the slope of this line so that it is not quite as vertical, you should expand the horizontal scale and compress the vertical scale. We will therefore set the horizontal scale (xScl) in the viewing window to a smaller number and the vertical scale (yScl) to a larger number. We are in effect rotating the line or stretching it away from the vertical axis.

The slope of the line in Example 3 is $m = \dfrac{\text{rise}}{\text{run}} = -110 = \dfrac{-110}{1}$. If we were to divide both the numerator and the denominator by 10 we would have $\dfrac{\text{rise}}{\text{run}} = \dfrac{-110}{1} = \dfrac{-11}{.1} = \dfrac{\text{yScl}}{\text{xScl}}$. To make things a little nicer, I will use $\dfrac{\text{yScl}}{\text{xScl}} = \dfrac{-10}{.1}$. So now set the RANGE window as follows:

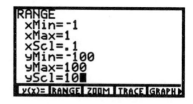
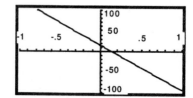

We can now see from the graph that the line crosses the x-axis somewhere between 0

and .2. We also know, from the given information, that the p-intercept is 15. We can now change our viewing window to be the following and obtain this graph:

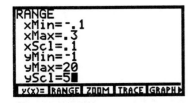
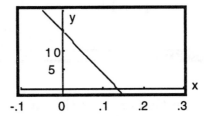

This graph leaves the impression of "steepness" intact, but makes such things as intersections with other lines and the axis something usable instead of just a blur of two nearly vertical lines colliding at some indistinct point.

Example 4: Graph $y = \dfrac{7}{327} q + \dfrac{23}{327}$

Solution: The $m = \dfrac{\text{rise}}{\text{run}} = \dfrac{7}{327}$, the y-intercept is $\dfrac{23}{327}$ (a number "close" to 0), and the graph on a ZSTD window sits on the x-axis.

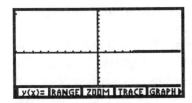

We need to rotate or stretch the line away from the x-axis; therefore we expand the y-axis and compress the x-axis. To do this look at $\dfrac{\text{rise}}{\text{run}} = \dfrac{7}{327} = \dfrac{\frac{7}{7}}{\frac{327}{7}} = \dfrac{1}{\frac{327}{7}} \doteq \dfrac{1}{47} = \dfrac{\text{yScl}}{\text{xScl}}$. To make things a little nicer, I will use $\dfrac{\text{yScl}}{\text{xScl}} = \dfrac{1}{50}$. So now set the RANGE window as follows:

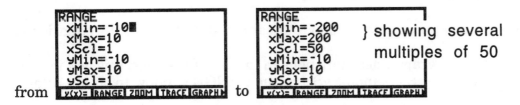

and if we need to see the graph mainly in the 1st quadrant:

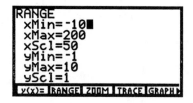
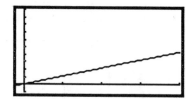

Caution: With the definition of slope that we are using we have to be careful when talking about a line in the usual geometric sense which has a vertical slope. For example, consider the line determined by the points (1,4) and (1,5). This is a vertical line with equation x=1:

$$\frac{5-4}{1-1} = \frac{5}{0}$$

You must not close your mind at this point just because we have an undefined result. As we shall see, vertical slopes and undefined fractions have tremendous significance in calculus and we will learn to deal with them whenever possible.

You have no doubt seen and dealt with equations of straight lines before now. We will take a little different view of lines than in the past. While lines have much importance on their own, many situations can only be dealt with by considering a collection of line segments, such as a graph of the stock market.

In addition, most curves do not change abruptly very often. We will want to look at what happens to a curve if you only change the domain values slightly. This is like putting a magnifying glass on a small piece of the curve. In most cases, this will result in a nearly linear relationship. Since straight lines are generally easier to deal with than curves, whenever possible we will use a straight line segment to deal with such small changes.

On the other hand, since we have calculator power available, we will not be forced to use straight lines in situations where they are inappropriate.

For a different approach to graphing example 3 and example 4 in this section, see Calculator II part B handout.

Information needed to work exercise 2.1

1. An ability to use the appropriate form of a linear equation to write the equation of a line.
2. An ability to enter and graph linear functions in the GRAPH editor; this will include the ability to adjust the xSCL and/or yScl in the RANGE window to obtain a "nice graph".

Exercises 2.1:

1. Given a linear price-quantity relationship such that when the quantity varies from 400 to 250 the price varies from $17,000 to $31,000, find:

   a. the slope
   b. the equation in point-slope form
   c. the equation in slope-intercept form.

2. Graph #1 above (GRAPH editor). Remember to adjust the viewing window to obtain a "nice graph".

3. Enter the points of #1 above in the STAT editor. Find the linear regression equation. Does this agree with your answer in #1c?

4. Graph $y = \dfrac{700q + 23}{3}$. Adjust the RANGE window and set scales so that you will have a "nice" graph.

5. A well known fact is that when the quantity of the comodity increases the price generally decreases. A certain price-quantity equation is known to be linear with slope -0.001 when 4000 units are available at $5 each. How many units would need to be available to lower the price to $4.00? (answer: q=5000, can you get it?)

Section 2.2 Demand, Supply, And Market Equilibrium

When a product is available in a free market economy, the price a consumer is willing to pay affects the quantity sold. Theoretically, as the price goes down, the quantity sold will increase. Conversely, if the price goes up, the quantity sold will decrease. We can see that the ratio $\frac{\Delta p}{\Delta q}$, change in price over change in quantity, will be negative for these functions. This is called the **law of demand**.

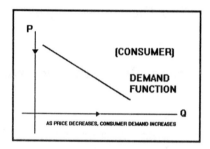

Associated with this law is the **law of supply**. As the price of an item increases, suppliers are willing to provide more, but as the price decrease they will supply less. Because of this direct relationship between increasing price - increasing supply, decreasing price - decreasing supply, the ratio $\frac{\Delta p}{\Delta q}$, change in price over change in quantity, will be positive.

You can see a fundamental conflict between what a supplier wants and what a consumer wants. Obviously, if the consumer got what they wanted, the price would be so low there would be no supply and the product would be unavailable. If the supplier got what they wanted the price would be so high almost no one would buy their product, thereby reducing their revenue. A happy medium between these two conflicting desires is called the **market equilibrium**. Market equilibrium occurs when the quantity of the commodity demanded equals the quantity supplied, **demand = supply**. This is indicated graphically by the intersection of the supply and demand curves.

When the word "supplier is used here, we are referring to the manufacturer or primary provider of the goods. The word "supplier" can also be applied to a retailer; but even with volume discounts, the retailer cannot generally buy goods for less than the manufacturer can produce them at a profit. As an example, consider any emerging technology such as DAT (Digital Audio Tape recording), which gives the user the ability to produce a digitally mastered tape with near perfect sound reproduction, matching that of a CD. What price are you willing to pay? In a recent survey of my classes, I found that most of you were willing to pay about $250-$300 tops. The supplier wants $750-$1200 and more. We see that the demand price is about $300, the supply price is around $900. Thus, you see the fundamental conflict between what a supplier wants and what a consumer wants.

Calculator Instructions II - Finding intersections of functions

Example 1: Equilibrium point

Use your calculator to find the equilibrium point in the following supply and demand equations. Graph each function and find the intersection.

$$\text{Demand: } p = -2q + 17, \quad \text{Supply: } p = 6q + 1$$

Solution:
1. Enter the demand equation as y1 and the supply equation as y2 in the GRAPH editor.
2. Set the Range window.
3. GRAPH
4. Find the intersection.

Your screen should be similar to the following:

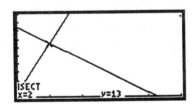   therefore q=2 and p=$13

You should consider this to be an unrealistic warmup example, for you will usually not be handed some neat equation to work with. It is more realistic to take marketplace observations. These usually occur in the form of points taken from observations. Such problems can strain your arithmetic a little, so let's work the following example on the calculator.

Example 2: Equilibrium point

Paul's ice cream parlor has discovered that at a price of .60 he can sell 300 scoops per day and at a price of .70 he can sell 250. He can buy from a number of different suppliers with different quality ice cream and different prices. Of the ones that sell ice cream up to his high standards, he has selected one company. They will reluctantly sell him what they have left over, up to the equivalent of 200 scoops per day at a price of .55, but will sell him as much as he wants at .65. Find the equilibrium price and quantity for this ice cream.

Solution: We will first assume that the price-quantity relationship is linear. We have two sets of points so we can use the point-slope form of the equation of a straight line after finding the slope.

Demand: (300, .60), (250, .70); (remember to put the quantity first, then the price).

Supply: (200, .55), (300, .65), and I picked 300 because this is apparently the most he can sell at the reduced price.

Using the point-slope form of the equation of a straight line after finding the slope of $m = \dfrac{.7 - .6}{250 - 300} = .1/-50 = .002$ we get:

Demand: Using the point-slope form: $p - .6 = -.002(q - 300)$ or $p = 1.2 - .002q$.

Supply: $p - .55 = \dfrac{.65 - .55}{300 - 200}(q - 200)$. This reduces to $p = \dfrac{q + 350}{1000}$ or $p = .001q + .35$.

NOTE: The STAT editor (Linear Regression) can also be used to find the demand and supply equations. Then you can either graph and find the intersection, or use SIMULT (simultaneous equations solver) to find the solution. You might also use the "old fashion method" - solve manually, using some algebra skills.

"Old fashion method": Setting the supply = demand

$.001q + .35 = 1.2 - .002q$
$.003q = .85$ therefore $q = .85/.003 = 283.3...$

Substitute this value of q into the supply of the demand function to get $p = .63$
If you choose the graphing method, enter the demand and supply equations in the GRAPH editor, set the RANGE window (a good scaling problem), GRAPH, and find the intersection. Your graph should be similar to the following:

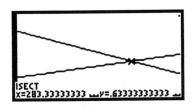 when the RANGE window is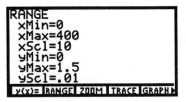

Market equilibrium : q=283.333 p=.63

Example 3: Back to reality

In actual practice, you are rarely (if ever) going to have anyone except your economics professor give you a demand or supply function. The marketplace determines these functions and you will have to come up with one based on market data. As a simple (if somewhat biased) example of this, let us right now try to determine the demand function for compact disks.

Imagine, if you will, that there is a new compact disk on the market by one of your favorite artists. You would really like to buy it, but given your current financial status, you check the price carefully before deciding what to do. In a recent sample of my class of 24 students, I asked for their reaction to the following prices. The question posed was: "At a price of ___ dollars, how many of you would buy this compact disk?" Student reaction is summarized in the table.

Price	30	25	20	15	10	5
Quantity	0	0	3	17	23	23

The total number of students willing to purchase came to 23 instead of 24 because one student said he would not purchase a CD at any price.

We now have a sample of consumer reaction to various prices. We would like to come up with a demand curve. We start by observing that at both the $30 to $25 price and at the $10 to $5 price there is no consumer reaction, so we discard these extremes. Doing a scatter plot on our calculator, we find that it is not terribly far from being a straight line. Thus, we decide to get the line of best fit for our data. One choice you have to make is to decide whether you want p (price) or q (quantity) to be the explicit variable (the variable the equation is solved for). Either will do since if you have an a linear equation it is very easy to solve explicitly in terms of the other variable.

For no other reason than that we have been getting p as a function of q so far, we do that here. You may easily make an argument that this is not the best choice since we really want to know customer reaction (q) to different prices (p). Here is a summary of our findings:

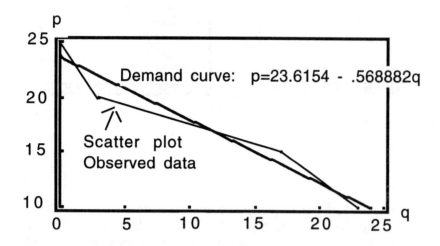

observed p	25	20	15	10
observed q	0	3	17	23
predicted p	23.6154	21.908754	13.944406	10.531114

Using our equation p=23.6154-.568882q or the FCST in the STAT editor on the calculator, let us find the reaction to a few pricing levels:

1. At a price of $16.95: 16.95=23.6154-0.568882q, we predict q=11.7
2. At a price of $12.95: 12.95=23.6154-0.568882q, we predict q=18.7

Information needed to work exercise 2.2

1. Equilibrium point: Demand=Supply **or** Demand-Supply=0

Exercises 2.2:

To the student:

The following exercises are so easy you will be tempted to do them by hand. Resist the temptation for one of the main purposes is to help you gain facility entering and solving equations on your calculator.

About rounding off:

Students are often puzzled about when they should round off an answer and when not to round off. Much of this confusion has been created by working problems out of context. If we are simply doing arithmetic, there generally is no good way to tell. However, when working within the context of a problem, common sense is the rule. For example, in problem 1 we are talking about units of oil. These units may very well be in 100,000 increments. To round off 3.742 to 3 would be changing an answer of 374,200 to 300,000. It should be clear that 74,200 barrels of oil should not be mathematically obliterated. Business students (heaven forbid) have also been known to round off 17.4912 to 17 when the units were millions of dollars. This removes $491,200 from future calculations. You do **NOT** have permission to do that to **MY** bank account! Unless you have had some advanced math classes in numerical methods, **you should not do any rounding off <u>while</u> you are working a problem**. Since you are working on a calculator, carrying extra digits as you work is not an inconvenience for you. In a few problems later on you will need to increase the number of digits of accuracy before you start working a problem to insure proper accuracy of the final answer and you will be warned about this beforehand.

1. At $32 a barrel, OPEC will produce all the oil you want up to 15 units. Due to political and economic considerations, at $17 a barrel they will produce only 8 units per day. On the other hand, consumers are willing to purchase all the oil they can get up to 12 units at $17 a barrel, but will only buy 9 units per day at $32 per barrel. What is the equilibrium point? (The answer is q = 10.8, p = $23. Do not round off the 10.8 for these are large units and 10.8 is a perfectly reasonable answer).

2. Sheet rock manufacturers are at the mercy of the marketplace like many suppliers for there are alternatives to using sheet rock in home building. If they raise the price too much, consumers will opt for different materials. On the other hand, if the price is low enough (compared to other materials) consumers will buy a great deal. Experience has taught that at a price of $12 per sheet, the demand falls to 3.8 units, but at a price of $3.50 per sheet the demand is 8 units. On the other hand, many manufacturers cannot make a profit at $3.50 so they will simply quit making it. This drops the available supply to 5.1 units. At $12 per sheet, the suppliers can make a killing so everyone jumps in and the amount available skyrockets to 10.5 units. What is the equilibrium point?

3. A department store chain will buy 200 television sets from a supplier if the price is $360 per set and 250 sets if the price is $290. The supplier will sell 120 sets for $275 per set, or 300 if the price is $360 per set. What is the equilibrium point?

4. A pet store can sell 10 bunnies a month if the selling price is $7.50, or 25 bunnies if the price is $4.50. The rabbit farm will supply 30 bunnies a month if the pet store pays $8.00 per bunny, or will supply 7 if the price is $4.00. What is the equilibrium

point of the pet bunny market?

5. An aerospace company will supply 50 advanced fighters for $8,500,000 each or 75 fighters for $10,000,000 each. A government will buy 60 fighters at $8,000,000 each or 70 at $7,000,000 each.

   a. Find the supply and demand functions.
   b. Find the equilibrium point.
   c. Get a graph with a reasonable appearance by adjusting your scale. Write down the scale you used.

6. A company will supply 1000 diskettes if they can get 39¢ for each disk or 250 if they get only 29¢ for each disk. A store can sell 800 if the price is 35¢ or 400 if the price is 45¢.

   a. What is the equilibrium point.
   b. Get a graph with a reasonable appearance by adjusting your scale. Write down the scale you used.

7. A hardware store will buy 200 light bulbs priced at 60¢ a piece or 150 at 80¢ a piece. The supplier will supply 175 bulbs at 75¢ a piece or 125 bulbs at 60¢ a piece.

   a. What is the equilibrium point.
   b. Get a graph with a reasonable appearance by adjusting your scale. Write down the scale you used.

8. The equilibrium point for an item is 1000 units at $1.50 per unit. However, when the price drops to $1.25 apiece, demand jumps to 1400 units, but there is a shortage of supply of 600 units. Assuming the underlying relationship is linear, find

   a. the supply and demand functions.
   b. what happens if the price jumps to $1.70

9. The market equilibrium for toasters is 560 units at $23.50. The plant is renovated with automated equipment, which cuts labor costs and allows the manufacturer to produce 75 more units at the same cost as before. Soon, the market equilibrium is 600 toasters at $19.95 each. Since this problem is a little tricky, look at the following hint.

   a. Find the original supply and demand functions.
   b. Find the new supply and demand functions.

   HINT: On this problem, consider this graph which shows some of the initial conditions. The result of being able to produce more for the same price means that the supply equation shifts downward, this will cause a resulting change in the equilibrium point. Now consider all the information you are given, including the coordinates of the new equilibrium point, (600, $19.95).

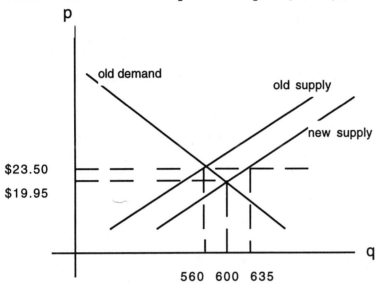

10. If the demand and supply functions for a textbook are $p+1.7q=89$ and $17p-11q=173$, respectively.

    a. Find the point of market equilibrium.
    b. If the price is 40, will there be a surplus or shortage of textbooks? What if the price is 30?
    c. Get a graph with a reasonable appearance by adjusting your scale. Write down the scale you used.

11. The demand function for pencils is $p+3q=78$. The supply function is $27p-18q=250$. When $p = 17$, are there too many or too few pencils to meet demand?

12. The following points were found for the demand and supply functions. Assuming the equations are linear, find the equilibrium point as well as possible.

q=	10	20	30	40	50	60
demand p	30		24		20	
supply p	11	15				27

13. The following points were found for the demand and supply functions. Assuming the equations are linear, find the equilibrium point as well as possible.

q=	100	150	200	250	300	500
demand p	3.00	2.50	2.40		2.00	
supply p	1.10	1.50	1.70			

14. It has been observed that at a price of $360, 200 TV sets are sold and at a price of $290 250 are sold. On the other hand, it has been noted that on the supply side,

number supplied	120	200	250	300	350
price per unit	$275	$320	$330	$360	$390

    a. Find the demand function.
    b. Find the supply function.
    c. Find the equilibrium point.

Section 2.3 Revenue, Cost, Profit And Break-Even Point

We must now look at the implications caused by the reaction of the market place to supply and demand. It should be easy for you to see that selling no sheetrock at $12 a sheet will not yield any revenue for the supplier and therefore there will be no profit. At the other extreme, selling 1 unit of sheet rock at a loss of $.25 per sheet is not good for profit either. Thus, in a sense we will examine more fully how the equilibrium point might be good for everyone concerned - both the supplier and the consumer. We therefore need to look at the following important terms and their interrelations.

IMPORTANT TERMS	SYMBOL or EQUATION	EXPLANATION
Units	q	Number of units
Demand Function	p = D(q)	p is the price per unit at which consumers are willing to buy q units
Supply Function	p = S(q)	p is the price per unit at which producers are willing to supply q units
Market Equilibrium Point	(q,p)	Point at which Demand = Supply
Fixed Costs	FC	A constant cost that does not depend on the number of units produced - rent, some utilities, insurance, etc.
Variable Cost per unit	vc	Costs that depend on the number of units produced - materials, labor, etc.
Total Variable Cost	VC = vc * q	Product of the variable cost per unit and the number of units
Total Cost	C = (vc * q) + FC or C = VC + FC	Sum of total variable costs and fixed costs
Revenue	R = p * q	Product of the selling price and the number of units (p in this equation is the demand price)
Profit	P = R - C	Difference between the revenue and the cost;
Break-even		The point where Cost = Revenue, Revenue - Cost = 0, Profit = 0

General Comments

The cost equation is a function of both the variable cost and the fixed cost. The fixed cost can include lights, security, rent, etc. Since it is difficult to get around fixed cost, this is often the first place a business can look when they wish to reduce expenses. For example, would a different lighting system reduce the electric bill? As an example, consider an item which costs $4 each for materials and the fixed cost per week is $1000. Your per week cost equation would be $C(q)=4q+1000$.

The demand function is the price the consumer is willing to pay. Note that price p and quantity q depend on each other. As the price decreases the quantity will generally increase causing the demand function to have a negative slope. Since p is the price the consumer is willing to pay, the revenue can be found if you know the demand function p. It will be $R = pq$, where p is the demand price and q is the number of units the consumers buy.

I want to point out to you a confusing notational situation that will arise in some printed material concerning demand and revenue. Since demand price, p, is a function of quantity q it is sometimes written as $p(q)$, "p of q", and, this being functional notation, the "of" DOES NOT MEAN TIMES. You are accustomed to this, but when an author writes revenue $R=pq$, you must realize that it can be written as $R(q)=pq$, where the "$R(q)$" is functional notation and "pq" means "p times q". A complete (and confusing) expansion can appear as "$R(q)=p(q)*q$", read "R of q equals p of q times q", combining both functional notation and multiplication on the same side. A good idea when you see such an expression is to ask your instructor for a careful explanation if you have the slightest doubt as to the meaning.

Example 1: Using demand and cost to get profit and break-even point

Suppose the selling price (demand price) for pencils is $.10. This means that the revenue = demand price x quantity will be $R(q) = .1q$, that is, if you sell 100 pencils your revenue will be $10. Suppose that your costs are as follows: fixed cost = $1000 and variable cost = $.05. Therefore, your cost equation $C(q) = .05q + 1000$ and the profit equation = revenue - cost is $P(q) = .1q - (.05q + 1000)$. To find the break-even point we use the definition of break-even:

Break-even point

The break-even point is the point where
Revenue = Cost
or
Revenue − Cost = 0

Since
Revenue − Cost = Profit,
the break-even point(s) can also be considered as the point(s) where:
Profit = 0

Simplifying the profit equation we get P(q) = .05q - 1000. To find the break-even point we set this equal to 0 and solve, getting q = 20,000. Note that since P = R - C and we have found the quantity that makes P = 0, R - C = 0, so R = C. This means that at the level of 20,000 units, revenue = cost, or R(20000) = C(20000). Thus we see that the break-even point is where the revenue finally catches up to the cost. In keeping with this concept, should P(19999) be positive or negative? What about P(20001)? Why?

Example 2: Graphing revenue, profit and cost

Suppose a company is making 5' concrete conduit sections. The cost equation is determined by the variable cost of each section of conduit, $73, plus the fixed cost of $1200. The equation is C(q) = 73q + 1200. In addition, suppose the sections can be sold for $110 each. This will yield a revenue function of R(q) = 110q. So:

Cost: C(q) = 73q + 1200

Revenue: R(q) = 110q

Profit: P(q) = R(q) − C(q)

P(q) = 110q − (73q + 1200)

P(q) = 37q - 1200

For this problem we will try to find the break-even point by first looking at the graph of the revenue and cost equations in the same plane. Graphing these equations with no axes adjustment will show only the revenue equation, R(q) = 110q since the cost equation crosses the y-axis at y = 1200. In addition to this problem, we find that the slope of the revenue equation, 110, is so steep that it is difficult to distinguish between the y-axis and the equation. We need to attack the two problems one at a time; we need to adjust the scaling.

Problem: getting both curves in the window

In this example, graphing both $C(q) = 73q + 1200$ and $R(q) = 110q$ in the same ZSTD window yields two nearly vertical lines; **not a desirable graph**.

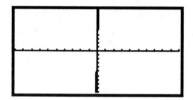

To get both curves to appear in the window simultaneously we need to do a little mental estimation. Since the cost equation has a slope of 73, and the revenue equation has a slope of 110, we need to pick some scale to flatten both curves out a bit, etc. rotate them away from the vertical axis. To do this we will use the ZFIT key in the ZOOM menu. We will set the xMin = -10, xMax = 50, xScl = 10 and ask the calculator to fit the horizontal scale accordingly. Making this adjustment we get the accompanying graph.

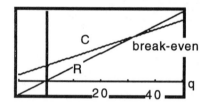

Checking your RANGE window, you should find that the calculator set yMin = -1100, yMax = 5500, and yScl = 100. To finish this example 2, we want to find the break-even point by finding the intersection of the revenue and the cost (break-even occurs where $R(q) = C(q)$). Using the ISECT key of the MATH menu in the GRAPH editor, you will find that break-even occurs at q=32.43 conduit sections.

Now let us look at solving this problem by graphing the profit function and finding the root. (Break-even occurs where P=0)

Graph $P = 37q - 1200$.

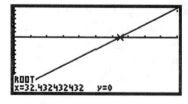

NOTE: This intersection (break-even) is easily found for linear equations by hand. It is not so easily done for non-linear equations; so you should practice on the calculator on these easy ones.

Information needed to work exercise 2.3

1. Cost = (variable cost) * (quantity) + (fixed cost)
2. Revenue = (price)*(quantity)
3. Profit = Revenue - Cost
4. Break-even point: Where Revenue=Cost, or Revenue-Cost=0, **OR** Profit=0

EXERCISE 2.3

1. A toy has a cost function C(x)=8x+2600 and a revenue function R(x)=24x.

   a. What is the profit function for the toy?
   b. What is the profit if 500 toys are sold?
   c. How many toys must be sold to break even.
   d. Graph the profit function and label the break even point.
   e. Find the break-even quantity by finding the intersection of the cost and the revenue functions. Does this agree with your answer in "d"?

2. Bill has a lemonade stand, and sells lemonade for $.50 a glass. He must pay Mr. Jones $3.00 a day to set up the stand in his yard, and each glass costs $.27 to make.

   a. What are the cost and revenue functions for Bill's lemonade?
   b. What is the break even point? (Use intersection of cost and revenue method)
   c. Graph the profit function and label the break even point. Does this agree with your answer in "b"?

3. A profit function is given by P(x)=4.5x-548.

   a. What are the slope and P intercept values of the function?
   b. Interpret the slope and P intercept values in words.

4. An item has a cost function C(x)=7.5x+237 and a revenue function R(x)=14x.

   a. What is the cost of producing 1 more item if 50 are currently being produced?
   b. What is the revenue received from selling 1 more item if 50 are currently being sold.
   c. Find the profit function.
   d. Find the profit made from producing and selling 1 more item after 50 have been produced and sold.
   e. Graph the profit function and label the break even point.

f. Find the break-even point by finding the intersection of the revenue and the cost functions. Does this agree with your answer in "e"?

5. Suppose the fixed costs for an new item will be $32,000, the variable costs are $3.50 per item, and the item will sell for $5.95. How many items must be produced and sold before the company recovers 50% of its initial (fixed) costs?

6. Given C(x)=4.12x+1080 and R(x)=14x, find:

   a. the profit function.
   b. the break even point. (Algebraically)

7. A linear cost function is given by C(x)=26x+354.

   a. What is the cost of producing 248 items?
   b. Find the fixed costs.
   c. Find the variable costs.

8. A farmer sells eggs for 60¢ a dozen. The cost of maintaining the hen-house is $1100 a month, and the cost of chicken feed averages 2¢ per egg produced.

   a. Find the cost and revenue functions.
   b. How many eggs must be laid to break even.
   c. Find the profit function.
   d. Graph the profit function and label the break even point.

9. A factory can produce a maximum of 400 shirts per day. Each shirt needs $7.50 in fabric, 21¢ in electricity, and $1.80 in labor to produce. The basic costs of operation of the factory are $380 a day. The shirts sell for $30.

   a. Find the cost function for the shirts.
   b. Find the profit function for the shirts.
   c. Find the break even point.
   d. Graph the profit function and label the break even point.

10. A new product will have variable costs of $28 per item and $325 a week in fixed costs. The factory can produce no more than 200 units a week.

    a. What price should be charged to make the break even point at 100 items per week.
    b. What would be the profit if the factory produced as many items as possible and the items were sold for the price in part a?

11. Susie sells sea shells by the sea shore. She sells the shells for 25¢ apiece. She can sell up to 80 shells a day.

   a. Find the revenue function R(x)
   b. What is the profit function?

12. A product sells for $8.00 per unit. The variable costs in the production of the item are $2.50 per unit, and the fixed costs are $25,000. If the company requires a 15% return on the fixed costs to break even:

   a. Find the Cost function.
   b. Find the break even point.
   d. Graph the profit function and label the break even point.

13. A commodity has cost function C(x)=15x+3200 and revenue function R(x)= 28x.

   a. Find the profit function for this item.
   b. What is the profit on 300 items?
   c. How many items must be sold to avoid losing money.

14. A supply function is given by $S(x) = 231 - \dfrac{1846.15}{x}$. A demand function is given by $D(x) = \dfrac{1583}{x+1}$. The corresponding cost function is C(x)=21x+138. Both functions are defined only for x≥8. Find:

   a. The equilibrium point.
   b. The break even point.
   c. Graph the profit function and label the break even point.

Section 2.4 Roots And Solving Equations

"Roots" in mathematics refers to those values of x where f(x) takes on the value 0. The determination of roots is often stated in a number of different ways and many different techniques can be employed to find them. You are often merely told to "solve" the equation for x. The word "solve" can become a point of confusion for some students. For example, to solve x+4=2 for x you would subtract 4 from both sides to get x=-2. However, to "solve" $x^2-1=0$ for x, you would factor and set each factor equal to zero: (x-1)(x+1)=0. The connection is that in solving x+4=2, the intermediate step of setting x+2=0 is not required. However, you are still finding a "root" of the equation.

In the last few sections we frequently took two equations such as p=-.2q+50 and p=q+5 and set them equal to find the equilibrium point, or, we set cost=revenue and solved to find the break even point. Do you recognize that you are simply finding the root of the difference of the two equations? Look at some of your graphs from the last section where you were asked to graph the break-even points after graphing the profit function. Did you note that the q of these points always occurred at the q where the profit function had the value 0?

Consider the following quadratic equation:

$x^2-300x+19950 = 0$

A quadratic equation is selected since we have a quadratic formula and can check the work we are about to do fairly easily. However, the procedure we are about to discuss will find the roots not only of a quadratic equation, but of any equation.

There are two numerical methods which are most frequently used. The simplest one is the **bisection method** which uses brute force and a huge number of calculations to find the answer. You can speed things up considerably by knowing about where the roots are. Any kind of decent graph will provide all the information you need to start the process.

The bisection method rests on a theorem that states that if one has a continuous function and $f(x_1)<0$ and $f(x_2)>0$, or vice versa, there must be a root between x_1 and x_2.

Pictorially, it looks like this:

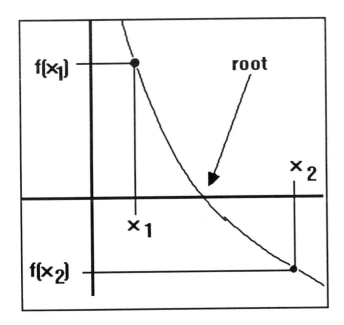

First we will need to graph $y = x^2 - 300x + 19950$. I used xMin = 0, xMax = 300, xScl = 50 and ZFIT for the y setup. Your graph should be as follows:

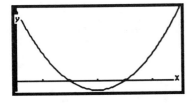

By inspection the first root lies close to 100; let us say between 80 and 120. The second root lies close to 200; let us say between 180 and 220.

The bisection method works by starting at either the left or right endpoint of the given interval, advancing in small increments and looking for the functional value to change signs. Suppose we first try to find the root that is in the interval (80, 120). The value for x_1 is 80 and the value of x_2 is 120. The process will begin as follows:

x	f(x)	
80	2350	
81	2211	
82	2074	
.		
.		
.		
99	51	
100	-50	a sign change occurs - therefore, bisect the interval (99, 100) to get 99.5

Now begin the process with this new interval: (99.5, 100).

x	f(x)	
99.5	.25	
100	-50	sign change occurs - therefore, bisect the interval (99.5, 100) to get 99.75

And again begin the process with the new interval: (99.75, 100).

x	f(x)	
99.75	-24.9375	
100	-50	NO sign change

We must now back up and bisect the interval (99.5, 99.75) to get 99.625. Our new interval will now be (99.5, 99.625).

x	f(x)	
99.5	.25	
99.625	-12.359375	sign change occurs - therefore bisect the interval to get 99.5625

Continue in this fashion, always halving the length of the interval and checking for a sign change from one end point to the other. If we specify accuracy to 6 decimal places, we find the approximate root to be 99.5039.

This procedure is not considered elegant by mathematicians, for some shooting in the dark is required and a great deal of brute force is used. However, when machines capable of doing millions of calculations per second are on everyone's desk, the procedure becomes very meritorious. Our graphics calculator provides us with several options for finding the root(s) of an equation. Let us begin with a very simple problem and explore our options.

Example 1:

Given the following cost and revenue function, find the break-even point by
  a. graphing the cost and the revenue functions - then finding their intersection,
  b. graphing the profit function (P = R - C) and find the root (x-intercept), i.e. the value of x where P = 0,
  c. use the SOLVER in the calculator. (See Calculator Handout III - Using The Solver)

$C(q) = 54q + 1200$ and $R(q) = 120q$

Solution:
  a. Enter the cost and revenue functions in the graphing editor of your calculator, set the RANGE window as xMin = 0 and xMax = 50 by xScl of 5 - then ZFIT the y values. Graph and use the ISECT feature of the MATH menu to find the intersection (break-even point). Your graph should look as follows:

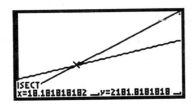

Your break-even point is q = 18.1818 and p = 2181.82.

b. If the revenue function was entered as y2 and the cost function entered as y1 (in part *a* above), then the profit function can be entered as y3 = y2 - y1. Deselect the y1 and y2 equations, and then graph the y3 equation using the same RANGE window. Your graph should look as follows:

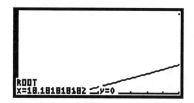

Now use the ROOT feature of the MATH menu of the graphing editor to find the root of the profit function x–intercept). Root: q=18.1818 and P=0. **NOTE** that the q's are the same for part a and b.

c. Enter either y2 = y1 or y3 = 0 in the SOLVER and solve for x. Your screen should be as follows:

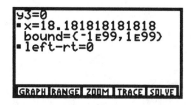

NOTE that you again obtain the same answer.

In this and similar problems, almost any reasonably scaled graph will show quickly where to look for a root. This is often not the case. As we shall shortly see, more complicated functions with many quick direction changes are difficult to deal with. In the next chapter we will learn some calculus and some techniques which will be useful in such circumstances.

Example 2:
Show that by solving $y_1=y_2$ we are finding roots of $y_1-y_2=0$. Use the graphing editor of your calculator to do this.

$y1 = -x^2 + 2x + 3$ and $y2 = x - 1$

Solution:
The intersection of $y_1= -x^2+2x+3$ and $y_2=x-1$ is the best way to think of the solution of $y1 = y2$, i.e. the points common to these two curves. Graphing both functions and using the ISECT feature twice, should give you the solution of:

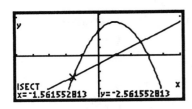 and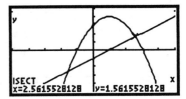

Now you need to turn the equality $y_1=y_2$ into a single equation equal to zero so that you can find the roots of $y_1 - y_2 = 0$.

Hence, starting with $y_1 = y_2$, subtract y_2 from both sides to get $y_1-y_2 = 0$. In our example, starting with $-x^2+2x+3 = x-1$, we would get

$-x^2+2x+3-(x-1) = 0$

Do I need to remind you how important the parenthesis are? The least carelessness on your part will lead to disaster here! You might also enter this in the graphing editor by : $y_3 = y_1-y_2$ to obtain a graph as follows:

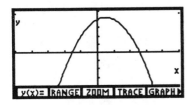

Now use the **ROOT** feature twice to find the roots (x-intercepts) of $y_1 - y_2 = 0$.

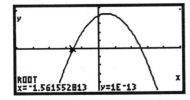 and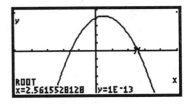

NOTE: the x values of the points of intersection are the same as the roots.

For a graphical comparison of what we have just done, look at the accompanying graph. As you can see, The x value of the points of intersection of the curves $y_1 = -x^2+2x+3$ and $y_2 = x-1$ are the same as the roots of $y_3 = y_1 - y_2$.

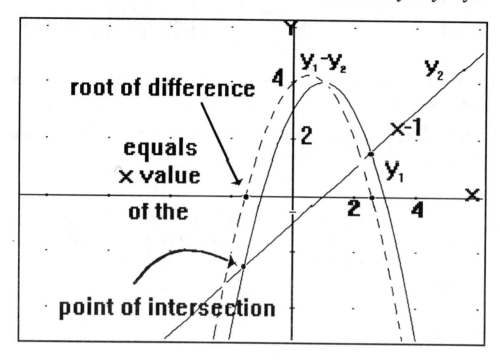

Example 3:

Suppose a demand curve is given by D: $p = \dfrac{200}{q+5}$, a supply curve by S: $p = \dfrac{80q - q^2}{100}$ and a cost equation $C(q) = 2q + 100$, where all equations are only valid for $q \leq 30$, ie., $0 \leq q \leq 30$. Find

a. the equilibrium point, and
b. the break-even point

Solution:

a. Enter all three equations into the graph editor. I will use y1 as the demand, y2 as the supply, and y3 as the cost. For now deselect the y3 (cost) equation. We will now graph y1 and y2 (Demand and Supply), use ISECT to find the point(s) of intersection. This point (s) will be the market equilibrium. Set the RANGE window as xMin = 0, xMax = 30, xScl = 2, and ZFIT the y's. Your graph should be as follows:

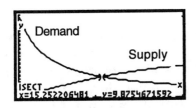

Market Equilibrium = (q,p) = (15.2522, 9.87547)

b. Recall that break-even occurs where Revenue = Cost or where Revenue - Cost = 0 (i.e. where Profit =0). We have the cost equation given and entered in the graph editor as y3; we therefore need the Revenue function. Revenue = Price * quantity; this is demand price, so R = D(q) * q.
We entered demand as y1, so enter the revenue as y4 = y1*q. Now graph y5 = y4 - y3 to get the graph of the profit function. Find the root(s) of the profit function to find the break-even point(s). Your graph should be as follows:

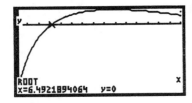

Break-even occurs where x=6.49219

You should note a very critical difference between the way you are accustomed to solving problems and this solution:

Most of our time was spent thinking about what needed to be done, using short hand notation to enter the facts (called functions or equations), then letting the calculator do all the actual work. As you become comfortable with this idea you will see that it is an extremely efficient way to work. Errors are tremendously reduced and you do not get all caught up in horrendous calculations when you should really be spending your time thinking about how to solve the problem. Notice what we did:

1. Solved D(q)=S(q) to get market equilibrium
2. Solved q*D(q)-C(q)=0 to get the break-even.

Our equations could have been at virtually any level of difficulty and we would still have spent about the same amount of time to arrive at the solution.

THIS IS THE POWER AVAILABLE TO YOU WHEN YOU USE <u>YOUR MIND</u> AND THE <u>CALCULATOR'S BRAWN</u>.

Example 4: New technology and demand functions

New technology frequently creates markets which did not previously exist. Two examples of this are television and tape recorders. In this case, when the new technology first hits the market, production costs are rather high, therefore producing a limited market. However, demand builds as more and more people wish they had this new technology. As the price drops demand increases and at some point the price of the technology reaches the "pocket book" level for the average consumer. At this point, relatively small decreases in price can cause fairly large increases in purchases. Consider the following "new technology" data:

q=	20	30	40	60	80	150
demand p	800	575		400	300	195
supply p	150			200	250	650

Enter the data in the stat editor, find:

a. an equation (best fit regression curve) for the demand function,
b. the supply function using a linear regression,
c. the missing data in the table, and
d. market equilibrium by finding the roots of Demand - Supply = 0.

Solution:

a. Entering the data, looking at the scatter plot, and checking the PWRR (Power Regression), you should find that the power function $p = 6235.0165441 \, x^{-.68811598}$ is a pretty good fit for the demand data. Store this in the graph editor as y1.

b. Entering the data and doing a linear regression, you should find the supply function to be $p = 1.408450704225 + 4.01408450742x$. Store this in the graph editor as y2.

c. Using the FCST (Forecast) feature in the STAT editor you should find the following for the missing data:

Supply - when q=30 then p=121.83 and when q=40 then p=161.97

Demand - when q=40 then p=492.53

d. Graphing y3 = y1 - y2 (i.e. y3 = Demand - Supply) and then using the ROOT feature of the MATH menu will give you the following:

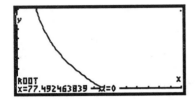

therefore market equilibrium occurs at q=77.49. Substituting this value in the demand or supply will give you p=312.48.

Information needed to work exercise 2.4

1. Be able to use the ISECT and ROOT feature of the MATH menu in the GRAPH editor to find the intersection(s) of two functions and the root of the difference of the two functions.
2. Be able to use the SOLVER editor to find intersection(s) of two functions or roots of equations after graphing to find approximate solution(s).

Exercise 2.4

In each problem, after you get the appropriate function, graph it and use the ISECT or ROOT feature of the MATH menu to find the intersection(s) or the required root(s).

1. A company has fixed costs of $31,000 and variable costs of x/2 + 250 dollars per item. Suppose the product's price is given by p = 1100 - 5/9 * x dollars per item. Find:

   (a) The profit function.
   (b) The break even point(s)

2. A good's supply and demand functions are given by 3p - q - 20 = 0 and (q+6)(q+p) = 2500 respectively. Find the point of market equilibrium.

3. Find the point of market equilibrium for the supply function $p = 2q^2 + 4q + 6$ and the demand function $p = 19 - 1.5q - q^2$.

4. A firm has variable costs of 3/5 x + 412 dollars per unit and fixed costs of $2000. The selling price is given by p = 2500 - 2/5 x.

   (a) Find the P(x).   (b) Find the break even points.

5. Find the points common to $f(q) = \dfrac{8q}{q+50}$ and $g(q) = \dfrac{3q+620}{q+20}$ $q \geq 0$.

6. Find the points common to $F(t) = 200e^{.08t}$ and $P(t) = 100(1 + \dfrac{.09}{4})^{4t}$ $t \geq 0$.

7. Find the point(s) common to $f(r) = e^{.08*20}$ and $g(r) = (1 + \dfrac{r}{4})^{80}$ $r \geq 0$

8-9. The following two (unrelated) sets of data were gathered from marketplace information.

You are to:

a. Make an estimate of the missing values after finding the appropriate functions.
b. Find the equilibrium point(s) by graphing the demand and the supply to find the intersection(s); then, graph y = demand - supply and find the root(s).
c. Find the revenue functions.

Data for problems 8:

q=	10	15	20	25	30	40
demand p	30	29			25	24
supply p	10		20		27	28

Use a quadratic regression for the demand and power regression for the supply in #8

Data for problems 9:

q=	10	20	30	40	50	60
demand p	30		24	20		17
supply p	10	19	23		21	

Use a quadratic regression for the demand and power regression for the supply in #9.

Section 2.5 Summary And Tracing To Get Maximums And Minimums

It is now time for you to put everything you have learned together into a comfortable package.

Business concepts:

Demand function - the relationship between the price consumers are willing to pay and the quantity available.

Supply function - the relationship between the price suppliers want and the amount they are willing to supply at a given price.

Revenue - Total income. It is usually found by multiplying price*quantity. It can be a function, often achieved by multiplying demand price*quantity, or just a number (20 items at $4 each=$80 revenue)

$$\text{Revenue} = \text{Demand price} * \text{quantity}$$
$$q*D(q)$$

Cost - Frequently represented by a function. Normally construct from variable cost and fixed cost like this:

Cost = (variable cost)*quantity + fixed cost

or by being constructed from total variable cost and fixed cost like this:

Cost = (total variable cost) + fixed cost

Note that the quantity is included in total variable cost so you do not multiply total variable cost by quantity.

Profit - Profit is calculated from revenue and cost. Profit=Revenue-Cost. But since Revenue is often calculated by using the demand function, the profit function often looks like this:

Profit = [demand]*quantity − (cost)

Equilibrium point - The equilibrium point is the point where the supply and the demand function have the same value. This can be found by either graphing both the supply and the demand functions and finding their intersection, or finding the roots of demand-supply equation.

$$\text{solve } D(q) - S(q) = 0$$

Break-even point(s) - The break-even point(s) occur when revenue = cost or when revenue − cost = 0. Since profit=revenue-cost then we can find break-even by either finding the intersection(s) of the revenue and the cost functions or by find the root(s) of the profit function.

**Profit = Revenue-Cost and Profit = 0
when Revenue − Cost = 0**

and this occurs when

**Revenue = Cost.
so: solve P(q) = 0**

As we have noted, the demand function is the price the consumer is willing to pay and p and q depend on each other. The demand function will generally have a negative slope. Since p is the price the consumer is willing to pay, the revenue can be found if you know the demand function p. It will be R = pq, where p is the demand price and q is the number of units the consumers buy.

Example 1: More about profit functions and break-even points

Suppose you have found the demand function to be p = -4q + 200. The revenue function will be R(q) = p*q = (-4q + 200)q. Note that this is a quadratic. Also note that whereas the demand function is simply a line with negative slope, the revenue function is a quadratic with a definite "peak". This peak would be the maximum revenue you could expect under the given conditions.

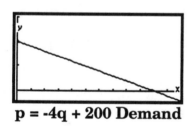
p = -4q + 200 Demand

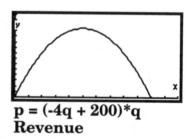
**p = (-4q + 200)*q
Revenue**

At first glance it might appear to you that knowing this peak would be ideal since it would maximize the revenue. That is true, but remember that a company can lose a great deal of money even with a large revenue, for it is the profit function that really matters. What we are really interested in is the peak of the profit function.

Suppose that in the above problem we know that the cost function C(q) has variable cost = $\sqrt{\dfrac{2000}{q}}$ and fixed cost $500. Our cost equation would then be

$$C(q) = \sqrt{\dfrac{2000}{q}} * q + 500 = \sqrt{\dfrac{2000}{q}} * \sqrt{\dfrac{q^2}{1}} + 500 = \sqrt{2000q} + 500.$$

We would then get a profit function:

$$P(q) = R(q) - C(q) = -4q^2 + 200q - (\sqrt{2000q} + 500).$$

Look at the graphs below.

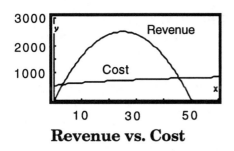
Revenue vs. Cost

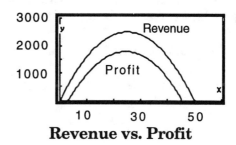
Revenue vs. Profit

The one on the left shows the plot of the cost function, a radical equation which appears quite linear due to the scale (**scale: (10,500)**), versus the revenue function which can clearly be seen to be a quadratic.

The graph on the right shows the profit function $R(q) - C(q)$. You can see that this subtraction will cause some small change in the peak of the profit function when compared to the peak of the revenue function. You should now try to estimate where each of these peaks are by using the TRACE menu in the GRAPH editor. However, the results are close together, and to solve the problem you really need to know some calculus. Using the knowledge you are going to gain in the next chapter, I have found the exact (to six digits) results to be:

the peak of the profit function occurs when $q = 24.7619$, whereas

the peak of the revenue function occurs when $q = 25.2381$

This difference is clearly caused by taking into account the effect of the cost function.

Example 2: Using least squares to find a demand function

A retailer observes the following relationship between quantity q of sales and prices p. (800,$12.00), (500,$14.00), (1400,$8.00), (100,$19.00), (1150, $11.00), (900, $11.50), (700, $12.50).
  a. Create the "most reasonable" retailers demand function.
  b. Create the revenue function from this demand function and give a reasonably close estimate of where the revenue is maximum.

Two points were observed on the suppliers reaction to pricing. At a price of $7.90 they

supplied 400 items and at a price of $9.00 they supplied 1700.
  c. Calculate the supply function.
  d. Find the equilibrium point.

The supplier has a variable cost of $6.80 and a fixed cost of $700. At a price of $7.90,
  e. Where is the break-even point?
  f. What is the quantity demanded at this price of $7.90?
  g. Calculate the suppliers profit at this sales and price level.

Solution for parts a and b:

The first step in creating a "reasonable" demand function is to see what a graph of the observed points looks like.

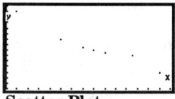
Scatter Plot

In this case it appears that the points nearly represent a linear function, so we will try to fit a linear function $p=a+bq$ to the given data. The exponential regression is a little better fit, but we will keep the demand function as simple as possible and still be "reasonable".

When we do this, we get the equation

$p = 18.668316831683 - .0076897689768977q$. Now store this as y1 in the GRAPH editor.

To generate the Revenue function from this demand function, we need only multiply the price p by the quantity to get pq. Enter the revenue function as y2 = y1*q.

At this point, we need to graph the revenue function and find where the greatest value is by tracing the graph. As we shall see shortly, calculus offers a better method for doing this. Note the scale of the graph must be changed since the revenue is price*quantity and these results will be around $9,000 to $12,000 on input quantities of around 1,000. Estimating the results we get about q = 1214 and $11,330.

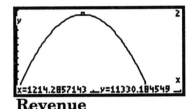
Revenue
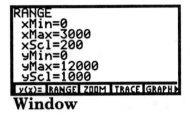
Window

Solution for parts c and d:
Since we have only two observed points we can use the two point form of the equation of a straight line yielding

$$p - 7.90 = \frac{9 - 7.90}{1700 - 400}(q - 400)$$
$$p = \frac{11q + 98300}{13000}$$

To find the equilibrium point you must then either equate this supply function to the demand function and solve, or subtract them and find the root. This could be done by graphing the supply and the demand to find the intersection(s) or by graphing the difference to find the root. Solving by either technique yields q = 1301.18, p = $8.66.

Solution to parts e, f, and g:
Since we are considering the profit function for the supplier in this part of the problem, the supplier's revenue function will be $R_s(q) = \$7.90q$ and the cost function will be $C_s(q) = 6.80q+700$. Hence:

$$P_s(q) = 7.9q - (6.8q+700) = 1.1q - 700$$

The break-even point is where revenue=cost or profit = 0. Solving for profit = 0 we get approximately q = 636. By setting the demand function equal to $7.90, we find demand quantity runs around 1400 units. At this level, the suppliers profit is $P_s(1400) = \$840$.

Example 3: Making an outline

At this point some of you will have fallen into the trap of sitting down with a calculator and entering "stuff". Your instructor can spot this if you ask questions like "What do I do with the 80?" If this is the case, your instructor is encouraged to give you a quiz like this:

© Saunders College Publishing

Sample quiz

In this quiz, you are NOT to provide any actual numerical answers. Rather, you are to set the problem up and explain what needs to be done to solve it.

It has been observed that at a price of $360, 200 TV sets are sold and at a price of $290, 250 are sold. On the other hand, it has been noted that on the supply side,

number supplied	120	200	250	300	350
price per unit	$275	$320	$330	$360	$390

Explain in steps exactly how you would find the equilibrium point. You should not do any arithmetic or algebra.

Answers:
1. Use the point-slope form of the equation of a line to find the demand function or use linear regression in the STAT editor and store this regression equation in the GRAPH editor.
2. Graph the data points (SCAT) and fit an appropriate least squares supply function.
3. Solve by graphing D(q)-S(q)=0 to find the root(s) or by graphing supply and demand to find the intersection(s).
4. If you use the root method, put this q value back in either equation (supply or demand) to get the price p.

Now state how you would find the revenue function

1. q*D(q)

Exercise 2.5

In these exercises, use your head and do some mental checking at each stage that you can. For example, if we are talking about a sales level of 5000 on an item that sells for about $10, demand and supply quantities should be in the 5,000 range, equilibrium price should be in the $10 range, and revenue should be in the $50,000 range. If answers are unreasonable, check your work. Such reasonableness checking is a measure of your understanding of what you are doing.

1. The following relationships for retailers were observed between quantity q of sales and price p: (1800,$1.00), (1500,$1.10), (1400,$1.20), (1200,$1.30), (1150, $1.40), (800, $1.50), (700, $1.60).

    a. Create the "most reasonable" retailers demand function.
    b. Create the revenue function from this demand function and give a reasonably close estimate of where the revenue is maximum. Hand draw a copy of your graph and label about where that maximum is (both coordinates).

2. In problem 1, only two points were observed on the suppliers reaction to pricing. At a price of $.90 they supplied 400 items and at a price of $1.20 they supplied 1600.

    a. Calculate the supply function.
    b. Using the equation you got in #1 for demand, find the equilibrium point.

3. In problem 2, the supplier has a variable cost of $.80 and a fixed cost of $150. At a demand price of $.90,

    a. Where is the break-even point?
    b. Given the quantities demanded in problem 1, justify the suppliers cut-back in supplies at a price of $.90.

4. Envy the supplier with a monopoly. Higher demand can be answered with higher prices. Suppose the marginal revenue is a constant $7 (ie., revenue = 7q). At $7 the supplier has sales (to retailers) of 12,000 items. To encourage sales, the price to the retailers is reduced to $5. Sales rise to 15,000 units. The supplier now raises the price to $8.

    a. Use this given data to generate the supplier's demand equation.
    b. Give the supplier's revenue equation.
    c. Using the equation for this revenue function, what would you predict sales to be at the $8 price? What would the suppliers revenue be at that price? You may wish to give several possible answers here and discuss your reasoning with me.

d. Graph the revenue function and give a reasonably close estimate of where the revenue is maximum. Hand draw a copy of your graph and label about where that maximum is (both coordinates).
e. Why would a supplier cut prices? In many cases, the retailer does not pass along the savings to the consumer. How can the supplier control the situation so the consumer does get the savings?

5. In problem 4, the supplier has a variable cost of $2 and a fixed cost of $12,000. Write down the cost function and

   a. Calculate the break-even point(s).
   b. Graph the Profit equation and label the break-even point(s).

6. It is frequently the case that we wish to predict long-term sales of some product based on initial consumer reaction. Test marketing of a new product is one example. In many diverse fields, the curve of best fit can be determined from past experience. We will use this to solve a recording industry problem.

   A CD recording and production studio has a weekly variable cost of $3 and a one-time fixed cost of $290,000 in the manufacture of new CD's. They observe the following end of week sales on a newly introduced CD: (week 0, 0), (week 1, 23,000), (week 2, 37,000), (week 3, 31,000), (week 4, 20,000).

   a. Assuming CD sales for a particular recording follow a quadratic curve, find the best curve relating week and sales.
   b. If the supplier charges $9 for each CD, give the revenue values for each week for the supplier.
   c. During what week is weekly revenue a maximum and about what is the weekly revenue at that time? Hand draw a graph of the weekly revenue function and label this maximum (both coordinates).
   d. Find the total revenue function $R(n)=9(y(1)+y(2)+...+y(n-1)+y(n))$ where $y(x)$ is the end of week function you found in part a.
   e. Find the total cost function and the break-even point(s).
   f. Find and graph the toal profit function. During what week should production be discontinued? Any answer you give must be justified with words and a graph.

7. Suppose we have a cost function $C(q) = 7 + 5q + .1q^2$. Let the revenue function be $R(q) = 10q - .1q^2$.

   a. Find the profit function.
   b. Graph the revenue function, the cost function, and the profit function all on the same axes.
   c. About where is the peak of the profit function?
   d. Would q be greater for the peak of the profit function or the peak of the revenue function?

8. At a price of $49.95, 123,000 Game Genies were sold. At a price of $54.95, 120,000 were sold, and at a price of $59.95, 90,000 were sold.

   a. Find and justify a demand function.
   b. The supplier has a monopoly. No matter how many you buy the price is $37. Do you have a best strategy? What is it?

9. Envy the Game Genie supplier with a monopoly. Higher demand can be answered with higher prices. The marginal revenue is a constant at $37 (revenue = 37q). At $37 the supplier has sales of 300,000 items. To encourage sales the price is cut so that the marginal revenue is reduced to $30. Sales rise to 315,000 units. The supplier now raises the price to $42.

   a. Give the equation for the revenue function.
   b. Using the equation for the revenue function, what would you predict sales to be at the $42 price? What is the predicted revenue at this price?

CHAPTER 3: DIFFERENTIAL CALCULUS

Section 3.1 Introduction To Marginal Concept

There are many different ways to begin calculus. We will use the most easily understood, if not the most useful. Let's begin by looking again at the idea of slopes.

Marginal.....

An important concept in the world of business and economics is that of "marginal". We will look at marginal cost, marginal revenue and marginal profit. The general meaning of the word marginal "whatever" is the additional "whatever" resulting from increasing the quantity by one. The formal definitions are:

Marginal cost: The additional cost from producing one more item. If $C(q)$ is a cost function, the marginal cost of the $q+1^{st}$ item is $\mathbf{C(q+1) - C(q)}$.

Marginal revenue: The additional revenue gained by selling one more item. If $R(q)$ is a revenue function, the marginal revenue for the $q+1^{st}$ item is $\mathbf{R(q+1) - R(q)}$.

Marginal profit: The additional profit gained by the production and selling of one more item. If $P(q)$ is a profit function, the marginal profit for the $q+1^{st}$ item is $\mathbf{P(q+1) - P(q)}$.

Since profit = revenue − cost,

marginal profit = (marginal revenue) − (marginal cost).

Note in these definitions that the word "marginal" refers to a particular level of production or sales. These values can (and often will) change as production and sales increase or decrease.

The usual situation you will find when working a problem is that someone (you, your accountant, etc.) has collected some data by making observations or recording sales, prices, etc. This means that the data will occur as a collection of discrete points. To keep things as simple as possible at this introductory stage we will consider the following ordered pairs, but I want you to realize right now that this is an idealized situation to make it easier for you to understand the concepts involved.

For example, suppose we were recording the unit by unit revenue received from the sale of a product. This data would appear as some kind of ordered pair, (units,revenue).

Revenue table:

units	1	2	3	4
revenue	$3	$6	$9	$12

The marginal revenue at a particular sales level q is the revenue gained by looking at the sale of one more item. Hence the marginal revenue at the 2nd item is:

$$R(2)-R(1) = \$6.00 - \$3.00 = \$3.00.$$

In fact, from our data we see that the marginal revenue at each recorded level is $3.00. This is not unusual, it simply means that we are charging a fixed price of $3 for each item sold. This is a typical linear revenue function, $R(q) = 3q$. Notice that we started with discrete data but have been able to turn this into a function. What is the advantage of doing this?

Given the work you have already done you should know that once you have a function, predictions for various levels of sales become very easy (assuming you have the RIGHT function!). For example, if our function is accurate, we can quickly say that the revenue at sales = 101 is $R(101) = \$303$. It is useful in calculus to express results in terms of functions.

Ideally, most business calculus books pretend that this can always be done. Realistically, it is difficult if not impossible to do this. Hence, we will usually approach every problem from both points of view; part of the problem may have a function to deal with while another part may have only discrete points.

Using this approach, suppose we have, by some arcane and questionable technique, determined that accompanying our linear revenue data above we have the following cost equation:

$$C(q) = \frac{1}{75}q^2 + \frac{q}{6} + 75$$

Calculator Instructions IV-Using The Table Program For The TI-85

Using the cost values we add to the table to get

Revenue vs. Cost table:

units	1	2	3	4
revenue	$3	$6	$9	$12
cost	$75.18	$75.39	$75.62	$75.88

Comparing the two tables for revenue and cost, we see that cost is greater than revenue at all values: R(1) = $3, but C(1) = $75.18, etc., potentially a <u>very</u> depressing state of affairs.

Will the revenue ever catch up? Will we ever get rich in this business? When can we buy the Mercedes? Or go to Switzerland skiing? Or..............are we going to file for you-know-what?

The answer lies in examining the profit function carefully. Recall that
he answer lies in examining the profit function carefully. Recall that

Profit = Revenue – Cost

We first need to take a general view of the revenue and cost functions. Note that we are dealing with a linear revenue function and a quadratic cost function; in all likelihood we will have two points of intersection. We must adjust our RANGE window so that we not only have nice graphs, but we must show both intersections. For example, setting your RANGE window as follows will give you the following graph.

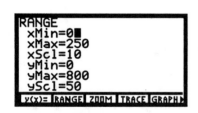
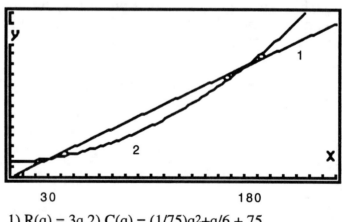

1) $R(q) = 3q$ 2) $C(q) = (1/75)q^2 + q/6 + 75$
Revenue vs. Cost

You should notice that at about q = 30 your revenue begins to exceed your cost, while at about q = 180 the situation reverses again. Clearly, if there is any way for us to make a go of this business we had better examine what happens in the interval (30, 180) VERY carefully.

The checking we need to do not only includes the obvious, which is the comparison of the revenue and the cost at various values, but also the less obvious view, that of looking for **trends** in the profit. In the problem at hand we are fortunate in having all of the data we need. What if we could only get values of the revenue and cost for q up to 100? In this case, we would like to see if there is any trend in the functions so that we will have some idea of what is going to happen next. As we shall see, the marginal cost, marginal revenue, and marginal profit will help us in analyzing some of these trends.

It appears that a good place to do some checking is near the 100th item. Calculating values for the revenue and cost within this interval we get the following table:

Revenue vs. cost table:

q units	101	102	103	104	105
$R(q) = 3q$	$303.00	$306.00	$309.00	$312.00	$315.00
$C(x) = \frac{1}{75}q^2 + \frac{q}{6} + 75$	227.85	230.72	233.62	236.55	239.50

Just as predicted by our original graph, in this interval the revenue exceeds the cost. Of equal importance is a comparison of the marginal revenue and the marginal cost. The marginal revenue is $3 everywhere, but the marginal cost for the 102nd item would be:

$$C(102) - C(101) = \$230.72 - \$227.85 = \$2.87$$

It is an easy task to interpret this result: we sold the 102nd item for more than it cost us. Our conclusion: This is good! Looking at the remaining marginal cost figures, we see that in each case the marginal cost is less than the marginal revenue of $3, and also note that the marginal cost seems to be steadily increasing at this level.

q	101	102	103	104	105
C	$227.85	$230.72	$233.62	$236.55	$239.50
MC		2.87	2.90	2.93	2.95

Unfortunately, we can see that the cost of production varies with the amount produced but the revenue does not. What does this mean at different production levels? Let's first compare the marginal cost function and the marginal revenue function at several points, while looking at the profit function.

Reviewing the formulas we will be using:

1. Since Revenue = $R(q) = 3q$, marginal revenue = $R(q+1) - R(q) = 3(q+1) - 3q = 3$.

$$MR(q+1) = R(q+1) - R(q) = 3$$

2. Since Cost=$C(q) = \frac{1}{75}q^2 + \frac{q}{6} + 75$, marginal cost = $C(q+1) - C(q)$

$$MC(q+1) = C(q+1) - C(q) = \left[\frac{1}{75}(q+1)^2 + \frac{(q+1)}{6} + 75\right] - (\frac{1}{75}q^2 + \frac{q}{6} + 75)$$

3. Since Profit = revenue − cost, $P(q) = 3q - (\frac{1}{75}q^2 + \frac{q}{6} + 75)$

units	2	36	101	151	201
revenue	$6	$108	$303	$453	$603
cost	$75.39	$98.28	$227.85	$404.18	$647.18
profit	-$69.39	$9.72	$75.15	$48.82	-$44.18

What a confusing mass of data! How can we hope to make any sense out of this? The answer lies in examining selected portions very carefully. First let us examine the profit while we look at the revenue vs. the cost carefully. Given only the numbers in our table, we see that everything looked fine until we hit the 201st item, at which point we were suddenly losing money again.

Was there any **PRIOR** warning that we would start losing money at about the 201st item? The answer is a resounding **YES**. We can see disaster approaching by looking at the marginal revenue vs. the marginal cost. To calculate the marginal values, we will first calculate the revenue and cost at the following values: (Use the EVAL menu in the GRAPH editor to fill in the table as follows)

q	1	2	35	36	100	101	150	151	200	201
R(q)	$3	$6	$105	$108	$300	$303	$450	$453	$600	$603
C(q)	75.18	75.39	97.17	98.28	225	227.85	400	404.18	641.67	647.18

After completing this table, calculate the marginal revenue, marginal cost, and the profit for each location by subtracting: $MR(2) = R(2) - R(1)$, etc. Note that it takes two revenues to make one marginal revenue, two costs to make one marginal cost, etc.

Now look at the marginal revenue vs. the marginal cost results and compare to the profit:

q	1 - 2	35 - 36	100 - 101	150 - 151	200 - 201
MR	$3	$3	$3	$3	$3
MC	$0.21	$1.11	$2.85	$4.18	$5.51
profit	-$69.39	$9.72	$75.15	$48.82	-$44.18

Since the marginal revenue is always $3, our task is relatively easy. We only need to look at the marginal cost. Over the range from 2 to 201 we see a slow but steady rise in the marginal cost. However, in the range from 2 to 101 the marginal cost is perpetually below that of the marginal revenue. This means that our total profit will gradually increase since we are constantly selling each item for more than the cost of production. A look at our profit function shows that this is indeed the case; **profit climbs** throughout the interval.

However, somewhere between the production of 101 and 151 items we see a rise in the marginal cost to the point where it passes the marginal revenue. This simply means that it costs more to produce the item than you are getting from sales. Note that at this point we are still making a profit ($48.82) but this cannot continue when it is costing more to produce an item than we are getting in sales. This will eventually force the profit to become negative and indeed this does happen somewhere between the 151st and 201st items.

Of an even more revealing nature are the figures for the marginal profit. First we calculate the marginal profit for the values given above. Since the profit function is

profit = revenue – cost

and you should already have these values, you are ready to calculate the profit values at each location, and then use these to calculate the marginal profit. However, a little algebra here will save you some work. Note that the marginal profit, **MP(q+1)**, for the (q+1)st item is:

$$MP(q+1) = P(q+1) - P(q)$$

Since we have already calculated the marginal revenue and the marginal cost, we would like to see if there is an easy relationship between these two and the marginal profit. If there is, we can quickly calculate these values.

MP(q+1) = P(q+1) − P(q), so consider P(q+1) and P(q).
      P(q+1) = R(q+1) − C(q+1) and
P(q) = R(q) − C(q)

Substituting the corresponding revenue and cost values for the profit equations and expanding our equation for the marginal profit we get:

$$MP(q+1) = \{R(q+1) - C(q+1)\} - \{R(q) - C(q)\}$$

and rearranging terms,

$$MP(q+1) = [R(q+1) - R(q)] - [C(q+1) - C(q)]$$

Now look at each term inside the brackets:

[R(q+1) − R(q)] = MR(q+1), and

[C(q+1) − C(q)] = MC(q+1).

Hence,
$$MP(q+1) = MR(q+1) - MC(q+1)$$

MARGINAL PROFIT = (MARGINAL REVENUE) − (MARGINAL COST)

$$MP = MR - MC$$

Calculating these values for marginal profit we get:

q	1 to 2	35 to 36	100-101	150-151	200-201
MR	$3	$3	$3	$3	$3
MC	$0.21	$1.11	$2.85	$4.18	$5.51
MP	$2.79	$1.89	$0.15	-$1.18	-$2.51
profit	$-69.39	$9.72	$75.15	$48.82	-$44.18

Looking at these marginal values, it only takes a moment to notice that the marginal profit seems to be a decreasing function throughout our range of interest. What does this mean?

Since the marginal profit is a measure of whether the profit is increasing or decreasing, looking at each individual entry tells us that profit is increasing **UNTIL** we get to about 100 items. At the 101st item marginal profit has decreased to a paltry 15¢. Shortly after this, the marginal profit becomes negative. The profit itself is still a positive $48.82, but it is clear this positive profit cannot continue and it doesn't. Profit becomes negative shortly thereafter.

Now let us look at our findings graphically and summarize our previous observations.

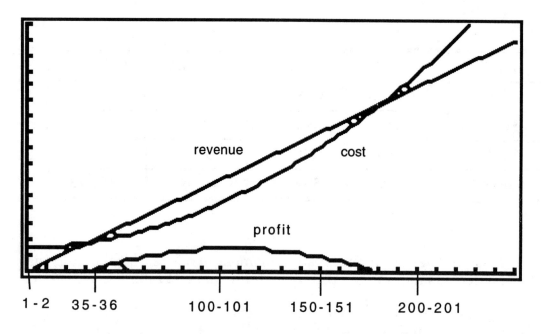

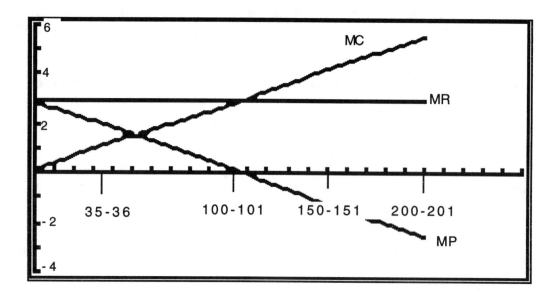

Observations:

1. R = 3q; MR = 3 at each recorded quantity since we are charging a fixed price of $3 for each item. Therefore, our marginal revenue looks like a horizontal line.

2. The graph of the profit function remained above the q-axis (i. e. P>0) as long as the graph of the revenue function remained above that of the cost function.

© Saunders College Publishing

3. At the intersection of MR and MC (i. e. where MR = MC), the MP = 0. **At this same quantity where MP = 0, the peak or maximum of the profit occurs.**

After you work a few problems like this and become comfortable with the interrelationships between a function and its "marginal relative" you will see how (in some cases) you can quickly calculate this marginal relative lurking in the shadows, ready to jump out and grab you at a moments notice, and use it to your own benefit!

Information needed to work exercises:

1. Cost=(total variable cost)+fixed cost
2. total variable cost=(variable cost)*quantity
3. Maximum profit occurs at the peak of the profit function which corresponds to the point where the marginal profit is 0.

Exercise 3.1

1. A product has total variable costs per week given by $VC(q) = 0.04q^2 - 12q$ and fixed costs of $1369 a week. The product sells for $4.32.

   a. Find the cost, revenue, and profit functions.
   b. Graph the revenue vs. the cost function vs. the profit functions.
   c. Fill in the following table: (use the EVAL menu in the GRAPH editor or the TABLE program to fill in the first three rows, use the third row to fill in the fourth row)

q	129	130	179	180	229	230	279	280
cost								
revenue								
profit								
MP	-	?	-	?	-	?	-	?

   d. Do an xyLine graph of the marginal profit in the STAT editor. About how many items should be produced per week to maximize the profit (i. e. about where does MP = 0)? Does your answer agree with the quantity at the peak of your profit function in part b?

© Saunders College Publishing

2. An item has a cost function $C(x) = 7.5x + 237$ and a revenue function $R(x) = 14x$, $0 \leq x \leq 200$

   a. Find the marginal cost and marginal revenue functions for the $(x+1)^{st}$ point.
   b. Would it make any difference where you calculated the marginal cost and marginal revenue functions?
   c. Find the profit function.
   d. Find the marginal profit function for the $(x+1)^{st}$ point.
   e. Find the marginal profit for the 51st unit.
   f. About how many items should be produced to maximize profit?

3. A study of an item's production costs gives the following data:

q	400	401	402	403	404	405
C(q)	$631.35	$631.74	$632.14	$632.54	$632.95	$633.85
R(q)						
P(q)						
MP		?	?	?	?	?

   The item sells for $2.00. Fill in the table. At a level of production of 402 units, what would be the best strategy?

4. A commodity has variable costs given by $vc = .24q^2 - 112.23q + 13405.23$ per unit and fixed costs $930. The item sells for $775.

   a. Find the revenue, cost and profit for each of the following quantities: q=190, q=210, q=230, q=250, q=270, q=280, q=290.
   b. Find the marginal cost, marginal profit and marginal revenue for each of the given quantities.
   c. Do an xyLine graph in the STAT editor of the marginal functions.
   d. Analyze the production strategy by looking at the xyline graphs of the marginal functions.
   e. Graph the cost, revenue, and profit functions in the GRAPH editor. Does your answer in "d" agree with your findings from these graphs?

Section 3.2 Graphing And Further Marginal Considerations

One of the main purposes of creating a graph is to gain some quick insight into a problem that may have much data involved. In fact, much of the study of statistics is devoted to compiling large amounts of data into a more compact form to make dissemination and understanding easier. In our case, our needs at first will be somewhat less ambitious. We wish to view the relationships between 2 or 3 sets of data and see what the inter relationships are. In doing this you will quickly find two fundamental problems:

1. Two data sets have dramatically different values. For example, one set of values is in the tens of thousands, the other in tens.

2. Changes of values within the data is very different. for example, one set of values ranges from 50 to 100, the other ranges from 70.01 to 72.5

The first problem is the most easily addressed. Two data sets with very disproportionate sized numbers will cause the following problem in graphing. Consider the following data:

Example 1: Shrinking and expanding the data range

quantity	15	16	17	18
Revenue	10,800	10805	10820	10840
MR		5	15	20

With the RANGE window set as follows you will get the accompanying graph.

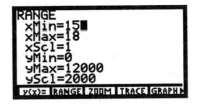 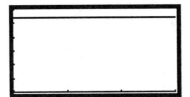

You can see that only the revenue data shows up. The effect of the marginal revenue is totally missing. This is the easiest type of problem to solve. Essentially, you must somehow cause the data sets to be superimposed, not widely separated. To do this, move the data physically closer by adding or subtracting appropriate amounts. Recall that we used this method in Chapter One. In this example we will subtract 10,780 from each revenue value giving the following result:

Quantity	15	16	17	18
Revenue	10,800	10,805	10,820	10,840
MR		5	15	20
Revenue-10,780	20	25	40	60

Placing the values that are in the bottom two rows onto a single scaled graph yields the accompanying graph.

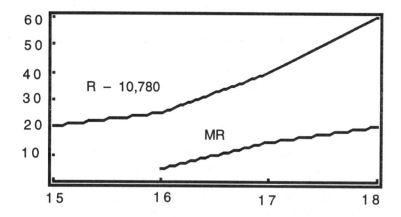

Since the marginal revenue is positive, the revenue is increasing. Since the marginal revenue is not only positive but also increasing, the revenue will not only be increasing but at an ever faster rate.

Thus, the first problem has been resolved:

1. Two data sets have dramatically different values. For example, one set of values is in the tens of thousands, the other in tens.

This problem is resolved by adding or subtracting appropriate values. Generally, when dealing with a function and its marginal values, this addition or subtraction is done to the original function and not to its marginal values.

Now we address the second problem:

2. Changes of values within the data are very different. For example, one set of values ranges from 50 to 100, the other ranges from 70.01 to 72.5.

Consider the following data set and graph:

Profit	$7,200	$17,198	$12,150	$20,050
Cost	$10,500	$10,550	$10,530	$10,700

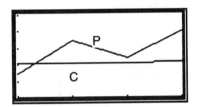 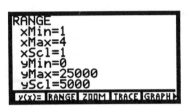

Notice that the values are fairly close together so they will both appear on the graph. Unfortunately, the changes in the cost are so small when compared to the profit changes that cost will appear only as a straight line. When this happens, no relationship between the two data sets will be revealed.

To correct this problem, you compare the changes in the two data sets by estimating the ranges in your head, if possible, since absolute accuracy here is not important. In the profit data, we see the range is about 13,000. In the cost data it is about 200. Dividing these numbers, $\frac{13,000}{200}$, we get 65. Thus, the profit changes about 65 times as much as the cost. Note carefully what we have found: $\frac{\text{total change in profit}}{\text{total change in cost}} = \frac{13,000}{200} = 65$. Thus, if we multiply the cost data by 65, we get $\frac{\text{profit}}{65 * \text{cost}} = 1$, ie., the relative change in both sets of values is now about the same order of magnitude.

Profit	$7,200	$17,198	$12,150	$20,050
Cost	$10,500	$10,550	$10,530	$10,700
cost*65	682,500.00	685,750.00	684,450.00	695,500.00

We now use our first procedure to get the data fairly close together. This can be done by subtracting 680,000 from the bottom line yielding the following data sets:

Profit	$7,200	$17,198	$12,150	$20,050
Cost	$10,500	$10,550	$10,530	$10,700
cost*65	682,500.00	685,750.00	684,450.00	695,500.00
-680,000	2,500.00	5,750.00	4,450.00	15,500.00

From the following graph you can see that cost follows profit to some extent, but to a much lesser degree. It is clear that dramatic changes in profit occur with little change in cost. What might your strategy be in such a situation?

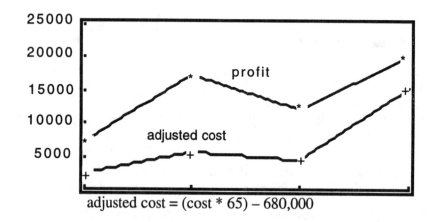

adjusted cost = (cost * 65) − 680,000

To summarize the general procedure in solving the problems that occur when changes in values within the data are very different:

1. Find the range of values for each function within the domain you wish to graph.

2. Divide one range by the other. Pick any close convenient positive number for the answer.

3. Multiply each value of the function you wish to adjust by the number you found in step 2.

4. You may then need to do some vertical adjustment of your data.

Example 2: Shrinking and expanding data

Now consider our problem from the last section again. This problem presents all of the graphing difficulties found above in one problem.

q	2	36	101	151	201
R	$6	$108	$303	$453	$603
C	$75.39	$98.28	$227.85	$404.18	$647.18
MR	$3	$3	$3	$3	$3
MC	$0.21	$1.11	$2.85	$4.18	$5.51
P	-$69.39	$9.72	$75.15	$48.82	-$44.18
MP	$2.79	$1.89	$0.15	-$1.18	-$2.51

Note the change in the profit from positive to negative as we pass through different sales levels. Profit can be positive or negative. Students sometimes feel uneasy when they get a negative answer for profit, but if there were no negative profit answers there would be no bankruptcies!

What we are concerned with at the moment is the change in the profit, the marginal profit. Ultimately, the profit function will reflect the influence of the marginal profit. For example, regardless of what the profit level is at a certain point in time, if the marginal profit is negative (you make less money for selling more items) and continues to have negative values (you make less **and** less money for selling more **and** more items), your profit function **MUST** be decreasing and eventually will become negative.

In this example, the profit function takes on both negative and positive values. Looking at the profit function, we would like to know if characteristics of the marginal profit function can be used to predict how the profit function will behave. The marginal profit reveals two very important trends for the profit function:

The **marginal profit** indicates whether the profit is heading up or down.

1. A **negative marginal profit** means that whatever your current profit level is your **profit is decreasing**. The current value of profit can be either positive or negative while decreasing.

2. A **positive marginal profit** means that whatever your current profit level is your **profit is increasing**. The trend of the profit is up. The current value of the profit can be positive or negative while increasing.

Of importance in our example is the continually decreasing trend of the marginal profit. It is often helpful to graph both the profit and marginal profit data on the same axis. By simply "eyeballing" such charts one can often note subtle tendencies. One difficulty in doing this is the fact that the orders of magnitude of the profit and marginal profit functions are often considerably different. One of our first concerns will be to figure out some simple process which will always yield an easy-to-read graph. Consider our current values for profit and marginal profit:

q	2	36	101	151	201
P	-$69.39	$9.72	$75.15	$48.82	-$44.18
MP	$2.79	$1.89	$0.15	-$1.18	-$2.51

The profit has values from $-69.39 to $75.15 while the marginal profit takes on values between $2.79 and -$2.51. Since you have extensive graphing experience, a moments reflection should convince you that if we graph the profit function to show a change of 150
(-75 to +75) and then put the marginal profit on the same graph, the tiny change of about 6
(-3 to +3) in the marginal profit will make its graph invisible.

In order to more dramatically illustrate what is happening it is important that the **relative change** in the data be on the same order of magnitude. To accomplish this you must distort the data by multiplication or division. Once this is done, relationships between the data will be numerically changed, but the graph may give a much more accurate picture of what is happening. It is difficult to give exact rules for doing this. Which do you distort and why? In our case, we want to see what effect the marginal profit has on the profit, so we would leave the profit values as they are and expand the marginal profit so that the magnitudes will match.

To do this we first look at the range of values taken on by each function within the interval we wish to graph. In our case this range is about 150 for profit (-69.39 to +75.15) and about 6 for marginal profit (+2.79 to -2.51). To get our multiplicative factor we divide 150/6 (approximately) and get 25 but select 30 to work with (you could also choose 20). Any close convenient number will do. We now take each marginal profit value and multiply by 30 yielding the following:

q	2	36	101	151	201
P	-$69.39	$9.72	$75.15	$48.82	-$44.18
MP	$2.79	$1.89	$0.15	-$1.18	-$2.51
MP*30	83.70	56.70	4.50	-35.40	-75.30

Note that the scales for the marginal profit and the profit are now roughly the same and can be graphed on the same axes. When you do this for general consumption, make sure your graph is clearly labeled to warn the unwary what you have done. Anyone who knows a little about business would get pretty excited if you told them the marginal profit was $83.70 instead of $2.79!

To summarize the general procedure:

1. **Find the range of values for each function within the domain you wish to graph. In our example, it was 150 and 6.**

2. **Divide one range by the other. Pick any close convenient positive number for the answer. In our example we looked at 150/6 and choose 30 as convenient.**

3. **Multiply each value of the function you wish to adjust by the number you found in step 2.**

At this point there is always the possibility that you might wish to add or subtract some value to the profit and marginal profit to get the final result conveniently close to the horizontal axis. Adding or subtracting values graphically slides the curves up or down on the scale. If we are only looking at trends, this will have no effect on our interpretation of what is happening. You may have noticed that graphs in the news media will often do this by creating a break in the vertical axis to indicate a jump in values.

Generally, when you are looking at trends, **you must not add or subtract values which would change the sign of the marginal profit**. Since positive marginal profit is **good** for the immediate future, and negative marginal profit is **bad** for the immediate future, the entire interpretation of the results is dramatically affected if you do this. Let's look at an example where some further adjustment is desirable.

Example 3: Scaling– one set of values is negative, the other positive

Consider the following data:

q	101	102	103	104	105
P	-16.18	-15.72	-15.29	-14.88	-14.50

With no adjustment in scaling, the graph will look like a straight line since the change in each value is relatively small (about 0.4) compared to the size of the numbers.

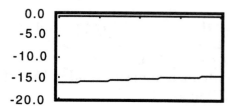

This is not the only problem. The small changes in the profit will result in small changes in the marginal profit - another scaling problem. We first calculate the marginal profit.

In this example, we have only discrete data. We have no formula to work with. The only values for profit we can get are those associated with sales from 101 to 105. We will be unable to generate any additional values. Recalling that the definition of marginal profit for the q+1st item is

$$MP(q+1) = P(q+1) - P(q)$$

you will realize that the entire marginal profit table cannot be filled in for there is no way to find the marginal profit at q = 101. Doing this would require a value at q=100 which we do not have. Hence our expaned table will look like this:

q	101	102	103	104	105
P	-16.18	-15.72	-15.29	-14.88	-14.50
MP		0.46	0.43	0.41	0.38

First, adjust the scale by using the multiplicative technique discussed earlier. The profit values vary by about 2, the marginal profit by .08.

1. The ranges are 2 and .08

2. 2/.08 is about 25

3. Multiply each marginal profit by 25.

q	101	102	103	104	105
P	-16.18	-15.72	-15.29	-14.88	-14.50
MP		0.46	0.43	0.41	0.38
MP*25		11.50	10.75	10.25	9.50

Graphing both the profit and the marginal profit times 25 (MP*25) on the same scale we see that they are relatively far apart causing the small changes in each to be considerably less noticeable. This would be much worse if they were even farther apart:

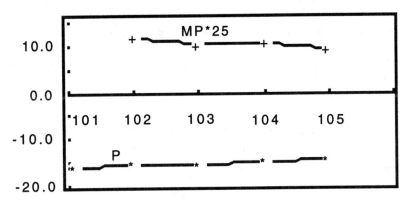

To reduce the physical separation between the profit and MP*25 you could either multiply the profit by -1 or add some positive number to each profit value. While the temptation to multiply P(q) by -1 is great because it's so easy, note what would happen to the marginal profit of this transformed equation, -P(q):

$$[-P(q+1)] - [-[P(q)] = P(q) - P(q+1)$$

Since MP = P(q+1) − P(q), P(q) − P(q+1) = -MP. **BAD NEWS!** You have just changed the sign of the marginal profit, in this case from a positive to a negative value which will lead to a total misinterpretation of what is happening. Instead of doing this we will add some appropriate value to the profit function.

It is important to note that in a transformation such as this, if you add (or subtract) the same number (say 25) from each profit entry, the marginal profit values for P(q)+k remain the same:

$$P(102) + 25 - [P(101)+25] = P(102) - P(101))$$

A reasonable way to proceed is to scoot the positive values that are relatively high in your graph down towards the horizontal axis making sure they remain positive, while sliding the negative values up towards the horizontal axis. This can be easily accomplished by viewing your data and/or your graph. In this case, look at the adjusted data set:

q	101	102	103	104	105
P	-16.18	-15.72	-15.29	-14.88	-14.50
MP		0.46	0.43	0.41	0.38
MP*25		11.50	10.75	10.25	9.50

Looking at the profit observe that the largest profit value is -14.50. In a graph, this value would be closest to the horizontal axis. We will slide our profit values upwards by adding 14 to each profit value.

Now look at the MP*25 values. 9.50 is the value closest to the horizontal axis so we will subtract 9 from each value. This slides all values of the MP*25 as close as possible to the horizontal axis **without changing their signs.**

units	101	102	103	104	105
profit	-16.18	-15.72	-15.29	-14.88	-14.50
Profit+14	-2.18	-1.72	-1.29	-.88	-.5
MP*25		11.50	10.75	10.25	9.50
MP*25-9		2.50	1.75	1.25	0.50

Plotting these curves we get the following:

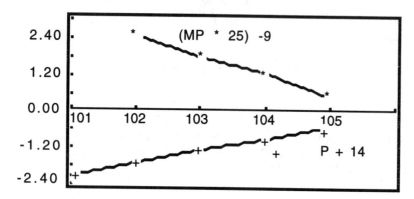

Looking at the figure we can draw several conclusions. The profit is increasing as the number of items increases, but at a slower rate. This is due to the fact that the marginal profit, while continuing to be positive, is on a downward track. If this trend were to continue, the marginal profit would eventually become zero, signalling no change in the profit, and then negative, signalling a decline in the profit.

Note that a marginal profit which is positive eventually yields a positive profit, since a positive marginal profit indicates the profit is rising. On the other hand, a marginal profit which is negative will eventually yield a negative profit since it indicates that the profit is on a downward trend.

Example 4: Average cost

Sometimes it is easiest to deal with the average cost of a large collection of items. For example, suppose the best you can do is to determine that the average cost for producing 0-1000 items is $8, and the average cost for the next 4000(1001 – 5000) is $7. Since average cost is calculated by taking you total cost and dividing by the number of items, the formula is: average cost = $\frac{cost}{quantity}$. Thus, given this data you can reconstruct the cost function, at least for these intervals. Multiplying both sides of the equation by quantity we get:

(average cost)*(quantity) = cost.

When constructing a table it may be desirable to fill in more values than just the end points. This can be done easily by simply choosing some in-between values:

quantity	200	600	1000	1001	2500	5000
Average cost	8	8	8	7	7	7
total cost	1600	4800	8000	7007	17500	35000

The quantity values have been chosen to cover the range of the data and also to show the change in cost at the endpoint = 1000. This technique can yield additional helpful information when graphing.

Information needed to work exercise 3.2

1. average cost = $\frac{cost}{quantity}$

Exercise 3.2

1. Two identical plants produce the same item, which sells for $4.25. Plant 1 produces about 71 items a day, the other produces about 80 items a day. Records show the following cost per day for each plant:

Plant 1

items	cost
70	285
71	276
72	269

Plant 2

items	cost
79	276
80	285
81	296

    a. Construct a table showing the revenue, cost, profit, and marginal profit for each plant.
    b. Graph the profit and the marginal profit on the same screen (adjusting when necessary) for each plant.
    c. Assuming both plants have the same cost function, what could be done to maximize profits for both?

2. A product sells for $9.65. Records indicate that the average cost of production of the item varies as follows:

# items	140	141	142	143	144	145
avg. cost	8.66	8.56	8.48	8.41	8.36	8.32
# items	146	147	148	149	150	151
avg. cost	8.30	8.29	8.30	8.32	8.35	8.39

    a. Find the cost, revenue, and profit at each level.
    b. Graph the cost, revenue and profit on the same axes.
    c. Find the marginal profit.
    d. Graph the marginal profit and the profit on the same axes.
    e. At a level of production of 148 units per day, what should be done to maximize profit?

3. Refer back to your work in problem 1 of exercise set 3.1. You should be almost ready to graph the profit and the marginal profit on the same axes. Do the necessary transformations to get an easy-to-read graph and produce the graph. (A product has total variable costs per week given by $VC(q) = .04q^2 - 12q$ and fixed costs of $1369 a week. The product sells for $4.32).

4. In this problem you are to create a table. Next, create the required graph.

   A study of an item's production costs gives the following data:

q	400	401	402	403	404	405	406	407	408
C	631.35	632.35	633.50	634.75	636.10	637.50	639.00	640.70	642.80

   If the item sells for $2.00,

   a. Enter the data given above and create the revenue.
   b. Find the profit and marginal profit for as many of the given values as you can.
   c. Graph both profit and marginal profit on the same axes.
   d. What would your best strategy be?

5. In this problem you are to create a table, then the required graphs.

   A study of an item's production costs gives the following data:

q	400	401	402	403	404	405	406	407	408
C	$631.35	632.35	633.50	634.75	636.10	637.50	639.00	640.40	641.70

   If the item sells for $2.00,

   a. Enter the data given above and create the revenue.
   b. Find the profit and marginal profit for as many of the given values as you can.
   c. Graph both profit and marginal profit on the same axes.
   d. What would your best strategy be?

6. Given the following cost values and a demand function p = 15 − .01q,

q	200	201	300	301	400	401	500	501
C	2440	2450	3440	3450	4360	4369	5200	5208

  a. Graph the revenue, cost, and profit on one set of axes.

  b. Find and graph the profit vs. the marginal profit over the given range and tell what a good strategy might be.

THE BICYCLE PROBLEM

A study shows that the monthly average costs in the production of a bicycle that sells for $199 vary with the number produced as follows:

bikes	avg. cost
0-150	200
151-299	172.0367
300-399	170.19
400+	165

PART I:

a. Calculate the cost, revenue, and profit values.
b. Graph the cost, revenue, and profit on the same axes.
c. Calculate the marginal profit.
d. Graph marginal profit vs. profit (on the same axes).

PART II:

Experience has shown us that bicycle sales are not flat and that we cannot sell all we can make. The demand for our product roughly follows the curve $q = 400e^{-.05(m-7)^2}$, where m=1,2,3,... is the month of the year. When entering this function in your calculator don't forget any signs or parentheses. The entry format would look like this:

$$400e^\wedge(-.05(m-7)^\wedge 2)$$

e. Calculate the demand q for our product for each month.
f. Calculate the revenue generated each month.
g. Calculate the profit generated each month.
h. Now what is your best strategy?

OPTIONAL PART III:

Bicycle sales actually have two peaks - one during bicycle season, a second less significant one around Christmas. The demand roughly follows the curve

$$q = 400e^{-.05(m-7)^2} + 100e^{-.05(m-11.5)^2},$$

where m=1,2,3,... is the month of the year.

i. Calculate the demand q for our product for each month.
j. Calculate the revenue generated each month.
k. Calculate the profit generated each month.
l. Now what is your best strategy?

Section 3.3 Introduction To The Derivative

Suppose we have cost and revenue functions as shown in this graph. We can see that since revenue exceeds cost from about a production level of 40 to a production level of 240, it is clear that in this region the profit is positive.

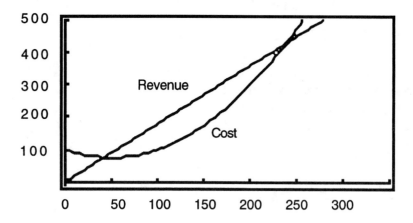

Constructing a graph of the profit function by subtracting, Revenue – Cost=Profit, we get the accompanying graph.

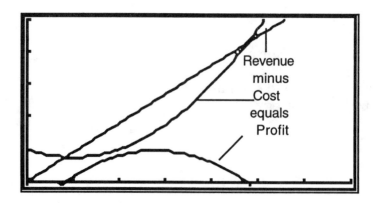

Taking a look at the profit function we can see that there is a region within which we can make a profit. We can also see that outside of this region our profit will be negative. We need to stay within some fairly small bounds to achieve a positive profit. Moreover, and most importantly in calculus, we see that the profit function has a definite peak, a point at which maximum profit occurs. How does one go about finding this peak where maximum profit occurs?

Let's sit back and look at the graph of the profit function. We can see that we will be most interested in finding the peak. How can we do this?

It is often useful when approaching a new problem like this to make observations about the characteristics of what we are seeing, much as a scientist makes observations about a new plant or animal. Given the profit curve in question, what sets the peak apart from any other part of the function?

1. It is the top or maximun of the function.

2. If one looks at something new, namely the tangent to the curve at a point and thinks for a while a brilliant idea might occur (it originally took about 200 years).

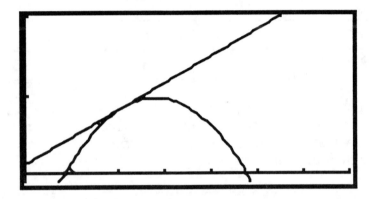

Instead of thinking in terms of the tangent line, think in terms of the **slope of the tangent line**. When you begin to do this you will soon come to an important observation:

The slope of the tangent line is 0 at the peak.

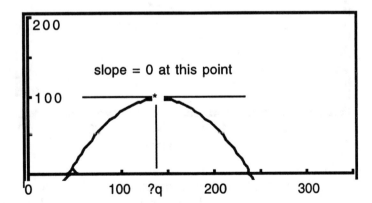

What good will this do you? Suppose you had some technique for finding an equation for the slope of the tangent line everywhere. The peak would occur at the **root** of that equation, and we can find roots. Is there such a procedure? The answer is yes.

If the data consisted of discrete observations we could find the slope between any two points and graph that, even though we might not be able to come up with an equation for such a slope "function".

As an example consider the following cost data:

units	30	50	70	101	102
cost	$76.67	$75	$83.67	$118.18	$119.72

When dealing with cost data our problem is the opposite from dealing with profit. In a profit situation we wish to know the maximum, which can occur at a peak. In dealing with cost data we wish to know the minimum which will occur at a valley. However, they share a single important concept:

Either a peak or a valley can occur when the slope of the tangent line=0

To illustrate how this would appear with the discrete cost data given above, look at the following graph:

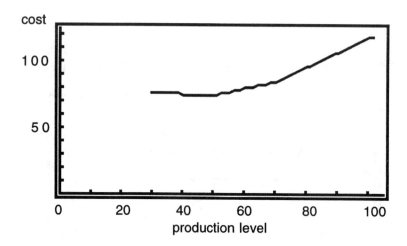

Even with our set of discrete points connected by line segments, it is easy to visualize that the minimum will occur at the apparent dip in the curve. Now that we have a possibly successful procedure to follow, let's try to:

1. find slopes, and
2. find slope functions.

This is where some careful observation and generalization will provide us with some new insights into the situation. Suppose now that we have cost data for the 101st to the 105th item as follows:

units	101	102	103	104	105
cost	$118.18	$119.72	$121.29	$122.88	$124.50
Marg cost		1.54	1.57	1.59	1.62

Let us look at the cost data from a slightly different viewpoint. Consider this graph.

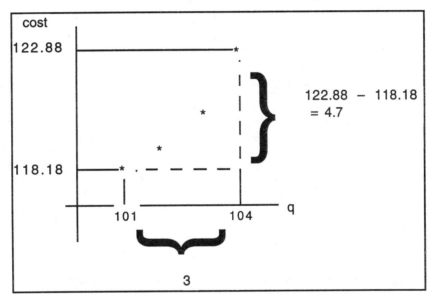

What is the average cost for the three items 101, 102, 103, and 104? To calculate this, we can make the following calculation:

$$\frac{C(104) - C(101)}{104 - 101} = \frac{122.88 - 118.18}{3} = \frac{4.70}{3} = 1.57$$

Recall that the marginal cost at the 102nd item was $1.54. It is clear that this average cost is close to that answer. How could we get closer?

Observation 1: The average cost we have just calculated is, in fact, the **slope** of the line joining the 101st point and the 104th point.

$$\frac{\text{rise}}{\text{run}} = \frac{C(104) - C(101)}{3}$$

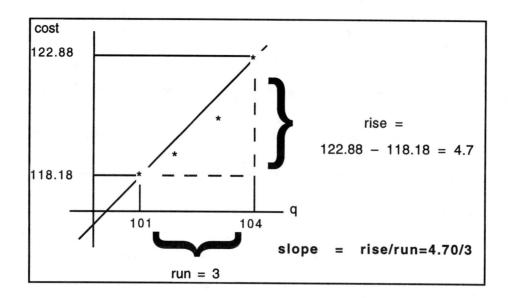

In fact, if we were to do the same calculation using the 101st point and the 103rd point, we would again get the slope. Viewing our initial calculation of the marginal cost of the 102nd item, C(102)-C(101) in this "slope" light, we see that it would be:

$$\frac{C(102) - C(101)}{1} = 1.54$$

Which is the slope of the line joining (101, C(101)) and (102, C(102))

Observation 2: Another way to view the marginal cost of the 102nd item is that it is the **slope** of the line joining the 101st item and the 102nd item.

This leads us to ask the following question:

If the process could be continued, and **if** the slope **at** the 101st item could be found, would this yield the marginal cost of the 101st item? Pictorially the process we are talking about looks like this.

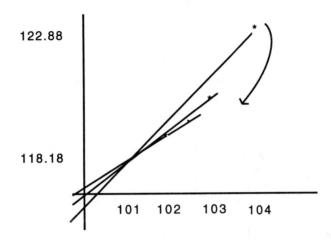

By letting our choice of points **move back towards the point at 102** we would be approaching the slope of the curve at q=101. In a business situation where the smallest increment you can have is one (1), you clearly cannot let the difference between two successive q values be less than one. However, it is often the case that we can construct a function from the observed data, just as we turned the point observations about the revenue into the function R(q)=3q in our example at the start of the chapter.

Viewing the situation in this light, there is no reason that we cannot let the difference between successive points get smaller and smaller. Viewing "marginal" as "slope", we would like to find the slope of the curve **at** some specified point, q. For example, in our cost function, if we could find the slope of the curve at q=101, this would be a close (but not exact) representation of the marginal cost of producing one more item, the 102nd. Hence the slope of the curve at q=101 would be nearly the fraction

$$\frac{C(102) - C(101)}{1}.$$

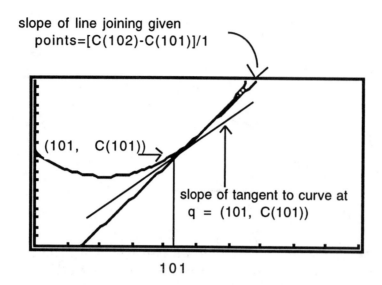

This brings us to a fundamental question which can motivate our introduction to differential calculus:

How does one find the slope of a curve **at** a point? We will answer this fundamental and most important question in the next section.

Exercise 3.3

1. Given the revenue function $R(q) = 3q$, find the slopes:

   $\dfrac{R(13)-R(10)}{3}$, $\dfrac{R(12)-R(10)}{2}$, and $\dfrac{R(11)-R(10)}{1}$. What do you conclude would **have** to be the slope **at** $q=10$? How does this correspond to your knowledge of linear functions?

2. Suppose you had an unknown profit function $P(q)$ and could observe that

   $\dfrac{P(104)-P(100)}{4} = .167$, $\dfrac{P(103)-P(100)}{3} = .164$, $\dfrac{P(102)-P(100)}{2} = .162$, and $\dfrac{P(101)-P(100)}{1} = .161$

   what would be a reasonable guess for the marginal profit **at** the sales level 100? Why do you say that?

3. Consider the function

$$MS(t,v) = v \frac{t^2 e^{\frac{-t^2}{2}}}{3!}, \quad t > 0,$$ t in years and v in thousands of sales. Look at the accompanying graph. This function is a type of curve which represents many situations. One is the marginal sales for retail items with a fixed life (such as text books, CD's, etc.). In the accompanying graph, a value for "v" has been chosen.

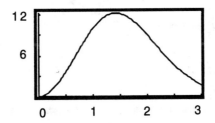

a. Where is the marginal sales the greatest? What is the numerical value of the marginal sales at that point?
b. In what range are marginal sales growing the fastest?
c. In what range are marginal sales falling the fastest?

Section 3.4 The Derivative

We now consider the problem raised in the last section. Given a function f(x) as indicated, how does one find the slope of the line tangent to the graph of f(x) **at** a given point?

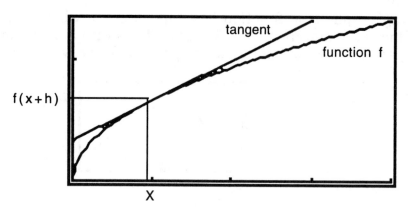

A little thought will lead you to the conclusion that no matter what is done, it doesn't seem possible that we can find the slope of the tangent line since our elementary knowledge of geometry tells us that we are not given enough information to find the equation of the line. No matter how we approach the problem, we are missing some crucial element needed to determine the line: Two points determine a line, but we only have one. One point and the slope will determine a line, but on a curve the slope changes all the time; hence, it seems we are in a quandary.

This is where the obvious approach to the problem will lead you astray. While it does seem entirely reasonable to assume that we need to know the equation of a line to find the slope of it, this is not the case. In our current problem, we could easily waste years trying to figure out how to get the line first, when, in fact, what we really want is the slope of that line.

To resolve this problem an entirely new concept is needed. Consider the accompanying illustration.

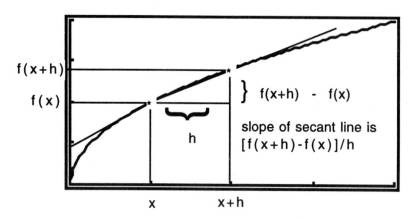

Given two points, we can calculate the slope of the line joining the two points by calculating the rise over the run. Notice that this IS NOT the answer to our question. This will give us the slope of a line joining the point in question, (x,f(x)) to another point, (x+h,f(x+h). What we really want is the slope of the tangent line **at** the point (x,f(x)).

However, consider the fact that with the quotient $\dfrac{f(x+h) - f(x)}{h}$ we can find the slope of a line joining **ANY** two points. If you don't see the significance of this, don't feel bad. The best mathematicians in the world did not see the significance of this for many years!

Since we are interested in the slope of the tangent to the curve at the point (x, f(x)), we start by looking at the slope of the line joining (x+h, f(x+h)), and the point we will ultimately be interested in: (x, f(x)).

This prepares us to do a little slight-of-hand. Suppose we consider what happens to this line as we let h decrease towards 0? The usual notation for this concept is written $\lim_{h \to 0}$, read "the limit as h tends to (or "approaches") zero". Since we are talking about the fraction representing the slope of the secant line, we would write $\lim_{h \to 0} \dfrac{f(x+h) - f(x)}{h}$, and you would read "the limit as h approaches zero of $\dfrac{f(x+h) - f(x)}{h}$". The quotient $\dfrac{f(x+h) - f(x)}{h}$ is sometimes called the **derivative quotient**.

For example, consider what happens if we take reductions in "h", say h_1, h_2, etc., reducing the size of "h" each time. Our secant line starts to rotate towards our tangent line. We would continue this process again, and again, and again......... You see that the **slope** of the secant line approaches the slope of the tangent line. This is the **most** important observation:

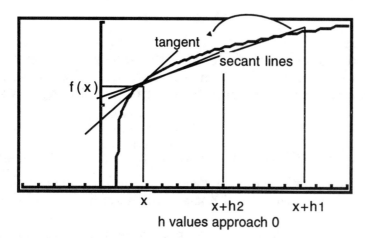

x   x+h2   x+h1
h values approach 0

If we have a function instead of a collection of discrete points, we can let h get smaller and smaller and still find the value f(x+h). Note that at each and every step of the way, we are simply calculating the slope of the line joining two points, so $\dfrac{f(x+h)-f(x)}{h}$ is valid every time.

Example 1: Letting the increment shrink

Let $f(x)=x^2+1$. Find f(x+2), f(x+1), f(x+.5), and f(x+h). Note here that we are interested in **ever decreasing** values for h in the expression f(x+h). We could let h increase, but since we wish to see what happens as the secant line approaches the tangent line, letting h increase would not serve our purpose.

a. $f(x+2) = (x+2)^2+1$
b. $f(x+1) = (x+1)^2+1$
c. $f(x+.5) = (x+.5)^2+1$
d. $f(x+h) = (x+h)^2+1$

Your prior training will make you want to jump right in and simplify these. As you will see in a moment, we will only be interested in simplifying the last one, f(x+h), and then not for the reason you probably have in mind.

While we will not wish to simplify those where h has a numerical value, we do wish to evaluate our slope fraction. However, there is a fly in the ointment. Note that as h gets closer and closer to 0, f(x+h) gets closer and closer to f(x), and, in fact, the numerator approaches 0; MEANWHILE, the denominator is approaching 0 at the same time. If we continue with this procedure, our slope fraction becomes $\dfrac{0}{0}$. This is undefined. What can be done about this undesirable state of affairs?

All is not lost - we haven't done our slight-of-hand. In order to see what can happen if we are clever, we need to take an example. In this case, the best example we can take is one for which we know the answer.

Example 2: Using the derivative quotient on a linear equation

Find the slope function for
$$f(x) = 2x+1$$

As you should already know, the slope of this straight line is 2 everywhere. Let us now attempt to use our formula to calculate the slope. If our reasoning when making up the slope equation to this point is correct, the answer should be 2. Since we need to fill in each part of the formula, we will do as much calculation as possible with our fraction before we even consider letting h approach 0:

$$f(x) = 2x + 1$$
$$f(x + h) = 2(x + h) + 1$$
$$= 2x + 2h + 1$$

Substituting these values into our equation in the appropriate place we get:

$$\frac{f(x+h) - f(x)}{h} = \frac{(2x + 2h + 1) - (2x + 1)}{h} = \frac{2h}{h} = 2$$

Notice that we did **as much algebra as possible** <u>before</u> taking the limit as h goes to 0. This is the slight-of-hand we can use with problems of this type. Now let h approach 0, ie., $\lim_{h \to 0} 2 = 2$. The number 2 is not affected by anything we do to h, so the result is 2. This answer is exactly what we were hoping for. This is a nice check, and a reassuring result but there is little value in using a fancy method to do something which can be done much faster and easier by more humble means.

Recall now the slope "picture" we had earlier. It will be useful in our current work to keep these slope values in the back of your mind.

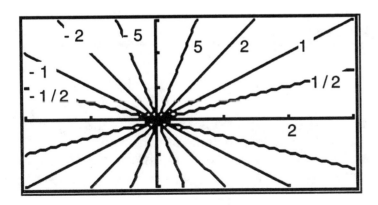

Let's look at a function which has a slope equation not as obvious as the one for f(x) = 2x+1.

Example 3: Using the derivative quotient on a quadratic equation

Consider the function f(x) = x². By looking at the curve and recalling what different slope values look like, we can guess three of the values the slope will take on:

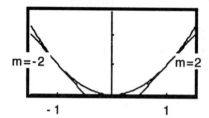

At -1, the slope is about -2,

At 0, the slope is exactly 0, and

At 1, the slope is about 2.

Now we will attempt to use the slope equation to find the exact values at these points. We again do as much calculation as we can before letting h approach 0:

$$f(x) = x^2$$
$$f(x+h) = (x + h)^2$$
$$= x^2 + 2xh + h^2$$

$$\frac{f(x+h) - f(x)}{h} = \frac{(x^2 + 2xh + h^2) - x^2}{h} = \frac{2xh + h^2}{h} = \frac{h(2x + h)}{h} = 2x + h$$

Notice again that we did **as much algebra as possible**. This is the slight-of-hand we used with the last example and can use with problems of this type. Having done all the algebraic manipulation we can, we again calculate. As h approaches 0, 2x + h will simply approach 2x + 0. Thus, using the slope equation for the function f(x) = x² yields the

slope equation for $f(x) = x^2$ is $2x$.

As a rough check of our answer, consider our guesses. Putting the values -1, 0, and 1 into the slope equation $2x$ yields: -2, 0, and 2, about what we expected.

Now let's try an example that is considerably more simple than either of the above. In fact, it is so simple that many students get it wrong!

Example 4: Using the derivative quotient on a constant function

What is the slope function for the function $f(x) = 5$? You should be able to see immediately that since the graph of $f(x) = 5$ is a horizontal line, the slope function will be 0.

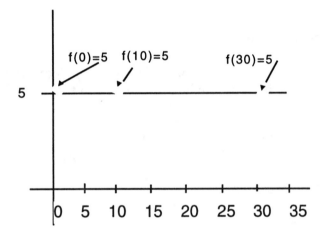

Calculating our slope equations we need only note that $f(x+h) = 5$ since the value of the function is 5 for any value of x you wish to pick.

$$\frac{f(x+h) - f(x)}{h} = \frac{5-5}{h} = \frac{0}{h} = 0$$

Note that I did the division while h **is not** zero. Now let h approach 0. This yields the value 0, which it should be since we are talking about the slope of the horizontal line $y = 5$.

Historically, due to the fact that the concept of the slope function was discovered by two different people (Newton in England and Leibnitz in Austria) on two different continents almost simultaneously, several different notations have evolved. The notation is also complicated by the fact that even within one course it is very convenient to use different notations depending on the circumstances.

NOTATION FOR SLOPE FUNCTIONS

Here is a partial listing:

slope function of f(x)

f'(x)

marginal f (Be careful here and recall our discussion at the end of section 3.2. For example, the application of the derivative to the profit function gives you the marginal profit **at** a point as opposed the difference P(q+1)-P(q) which is the marginal profit **for** the q+1st item. Further discussion of this difference will be presented in the next section.)

derivative of f with respect to x

rate of change of f with respect to x, or, as is more common in business, just "**rate**". Such as, if tax is 25% of your income or .25*I, the tax **rate** is .25

$\frac{df}{dx}$, read df dx or derivative of f with respect to x.

$\frac{d\,f(x)}{dx}$, read df of x with respect to x

f' is often used when there can be no confusion about the argument.

One of the common problems students have when first attempting to use the slope equation is in calculating f(x+h). I have found that a fail-safe way to do this is as follows:

Example 5: Substituting x+h for x

Consider the function $f(x) = x^2+5$. Think of the variable as being expressed by parenthesis, like this:

$f(\ ) = (\ )^2+5$.

The blanks are to be filled in with whatever you happen to be talking about at the moment - whether it be x, 5, or x+h. Thus, putting x+h in $f(\ ) = (\ )^2+5$ yields:

$$f(x+h) = (x+h)^2+5 = x^2+2xh+h^2$$

In exercise 7 you will see the equation f(x) = (x + 1)². Think of it like this:

$$f(\ ) = [(\ )+1]^2$$

so that f(x+h) = [(x+h)+1]². Expanded this gives:

$$(x+h+1)^2 = (x+h+1)(x+h+1) = x(x+h+1) + h(x+h+1) + 1(x+h+1) \text{ etc.}$$

Information needed to work Exercise 3.4:

1. $\lim\limits_{h \to 0} \dfrac{f(x+h) - f(x)}{h}$ read "the limit as h approaches zero of $\dfrac{f(x+h) - f(x)}{h}$".

EXERCISE 3.4

Use the definition of the slope function given above to find the slope function or $f'(x)$ for each of the following.

1. $f(x) = 5x+8$
2. $f(x) = 3x-1$
3. $f(x) = 2x^2+3$
4. $g(x) = 3x^2-5x+1$
5. $f(x) = \pi$
6. $g(x) = 2+\pi$
7. $f(x) = (x+1)^2$
8. $f(x) = (x+3)^2+4x-1$

Section 3.5 Finding Some Formulas And Theorems

We could calculate derivatives of each and every individual function we come across but this would be tedious and unnecessary. Let us calculate a range of simple derivatives and look for a general pattern. Hopefully, this will lead us to a general formula. We will call this formula "elementary derivative number 1".

EXERCISE

Calculate the slope function for the following functions. Use

$$f'(x) = \lim_{h \to 0} \frac{f(x+h) - f(x)}{h}$$

1. $f(x) = x^2$
2. $f(x) = x^3$
3. $f(x) = x^4$
4. $g(x) = x^{-1}$
5. $h(x) = x^{-2}$

Now make up the correct formula and write it in your notes. Do not write it in this book and spoil the fun for the next person to use it!

Elementary derivative number 1: (in your notes)

You may have learned to dread theorems when you had your high school math courses. This is most unfortunate. In this book we will reserve the word theorem only for things of extreme importance. In most cases, they will also be such powerful time-saving devices that you will be more than glad to memorize them without the urging of your instructor. We are about to make up two such theorems. We will first search for a pattern in the calculation of derivatives of functions of the form cf(x), where c is any constant. We will then search for a theorem concerning functions of the form $f \pm g$. What we really need is some way to deal with functions which are combinations of functions of the form x^n. For example, what does one do with $f(x) = 4x^3$? How about $h(x) = 4x^3 + 5x^2 - 3x + 1$? The answer can be found by looking at the definition of the derivative quotient. First note that $f(x) = 4x^3$ can be viewed as $4(x^3)$ so $f(x+h)$ is $4(x+h)^3$. This gives $\frac{4(x+h)^3 - 4x^3}{h} = 4 \frac{(x+h)^3 - x^3}{h}$, so the 4 can be factored outside before we take the limit and the first theorem we need becomes obvious:

Theorem 1: (constant times a function) $[cf(x)]' = c[f(x)]'$

To differentiate a constant times a function, take the constant out until you're done, then multiply by it.

Example: Calculating the derivative of $4x^3$, we view it as 4 times the derivative of x^3. Hence, the answer is $4(3x^2) = 12x^2$.

It should be clear that another theorem is needed, namely, how does one deal with linear combinations; expressions like $h(x) = 4x^3 + 5x^2 - 3x + 1$. Since we already have a theorem to deal with the constant multipliers, we only need a theorem to deal with the sum, $f + g$, or difference $f - g$ of two functions. Again, it is easiest to see what happens by looking at the definition for derivative, applying it to our current situation. Putting $f(x)$, $f(x+h)$, $g(x)$, and $g(x+h)$ into the definition we get the following easy result:

$$\frac{f(x+h) + g(x+h) - f(x) - g(x)}{h} = \frac{f(x+h) - f(x) + g(x+h) - g(x)}{h} = \frac{f(x+h) - f(x)}{h} + \frac{g(x+h) - g(x)}{h}$$

or, more simply remembered,

Theorem 2: (sum and difference) $[f \pm g]' = f' \pm g'$

The derivative of the sum (or difference) of two functions is the sum (or difference) of the derivatives.

Now you need a little "finger practice" in differentiating, so try the following.

Worked examples:

1. $x^3 - 4x^2$

solution: We first apply theorem 2, which says that one can take the derivative of the difference of two functions by differentiating each and subtracting the results. $\frac{d}{dx} x^3 = 3x^2$ and $\frac{d}{dx} 4x^2$ can be calculated using theorem 1.

$$\frac{d}{dx} 4x^2 = 4 \frac{d}{dx} x^2 = 4(2x) = 8x$$

Now that we have the results, theorem 2 tells us that the answer is $3x^2 - 8x$.

2. $11 + \dfrac{29}{x}$

solution: We will again use theorem 2, but before proceeding you **must** note that $1/x$ is not in the correct format for differentiating. Our fundamental derivative formula tells us how to differentiate x^n, and everything **must** be in this format. $1/x$ is not and you must use your knowledge of algebra to re-write it; $1/x = x^{-1}$.

$$\frac{d}{dx} 11 = \frac{d}{dx} 11x^0 = 0 \text{ and } \frac{d}{dx} \frac{29}{x} = 29 \frac{d}{dx} x^{-1} = 29(-x^{-2}) = -29x^{-2}$$

Now theorem 2 tells us that the answer is $0 - 29x^{-2} = -29x^{-2}$.

3. $\dfrac{x^3 - 5x^2 + 3x - 1}{x^2}$

solution: Since each term of the expression is not in the form cx^n, we must perform the algebraic operations necessary to get it into this form. First, break the fraction into its separate components:

$$\frac{x^3}{x^2} - \frac{5x^2}{x^2} + \frac{3x}{x^2} - \frac{1}{x^2}$$

Now reduce the terms and rewrite each term in the form x^n:

$$x - 5 + 3x^{-1} - x^{-2}$$

Theorem 2, Theorem 1, along with the rule for elementary derivative number 1 may be applied yielding:

$$1 - 0 - 3x^{-2} + 2x^{-3}$$

It is usually shown in more advanced courses that **elementary derivative number 1** holds true for any real exponent. That is, not only can you conclude that $\frac{d}{dx} x^n = nx^{n-1}$ when n is an integer, but the same holds true for any real number.

4. $x^{1/3} + \sqrt{x}$.

solution: $\frac{d}{dx} x^{1/3} = \frac{1}{3} x^{-2/3}$. To differentiate $\sqrt{x}$ we first rewrite it in exponential notation as $x^{1/2}$. Then, $\frac{d}{dx} x^{1/2} = \frac{1}{2} x^{-1/2}$.

Algebra needed in exercise 3.5:

1. $\frac{1}{x^n} = x^{-n}$, so $\frac{1}{x^{1/3}} = x^{-1/3}$

2. $\sqrt[5]{x^2} = x^{\frac{2}{5}}$

Exercise 3.5   Find $\frac{d}{dx}$ for each of the following.

1. $x^3 + 4x^2 + 5$
2. $99x^{100} + 57x^2 - \pi$
3. $4x^2 + 4x + \frac{4}{x} + \frac{4}{x^2}$
4. $\frac{4x^3 + x - 5}{x^2}$
5. $(3x + 1)^2$ (yes, you'll have to multiply it out!)
6. $(2x^3 + 1)(x + 1)$
7. $(4x + 1)^3$ (Ouch! I think I'm ready for another theorem.)
8. $5x^3 - \frac{3}{x} + \frac{5}{x^3}$
9. $\sqrt{x} + \sqrt[5]{x}$
10. $\frac{x^2 + 5x}{\sqrt[3]{x}}$

Section 3.6 Product And Quotient Theorems

We have learned how to differentiate functions which can be expressed as some linear combination of terms of the form x^n. What about the product and quotient of these functions? For example, without multiplying it out, how does one differentiate $(x^3+5x^2-3x+10)*(7x^4+x^2-12)$? Or, (and this is a little more difficult), how does one differentiate $\frac{3x+5}{7x-1}$? We need two more theorems, one for dealing with products, and one for dealing with quotients.

In developing the product theorem, "eternal optimism" would lead us to think that we could find $[fg]'$ by simply calculating $f'g'$. This would be nice, but let me show you that it won't work. Consider the function x^5. You should know by now that the derivative of this is $5x^4$. This function can be re-written as the product x^2*x^3. Calculating the derivative of each piece and multiplying we get $2x*3x^2=6x^3$. But the correct answer is $5x^4$ showing that our wishful thinking is only that. To do this correctly we need the product theorem:

Theorem 3 (product theorem): $[f*g]' = f'g + f g'$

To differentiate the product of two functions, take the derivative of one function and multiply by the other function, then add the result of doing the opposite.

This brings me to a warning that I will repeat several times for you:

In differentiating, there are no situations where two derivatives are taken in a single step.

As you shall see, from this point on we will need to use several theorems which require more than one differentiation to complete. However, when there are several steps, only one derivative will be taken in each step.

To see that Theorem 3 does in fact work, consider x^5 expressed as x^2*x^3 again. We will let $f=x^2$ and $g=x^3$. Then:

$[f*g]' = f'*g + f*g'$
$[x^2*x^3]' = [x^2]'*x^3 + x^2*[x^3]' =$
$(2x)*(x^3) + (x^2)*(3x^2) = 2x^4 + 3x^4 = 5x^4$

the correct answer. Note that you can execute this theorem in any order since only multiplication and addition are involved; $f'*g + f*g' = f*g' + f'*g$.

Example 1:

Differentiate $(x^3+5x^2-3x+10)*(7x^4+x^2-12)$.

solution: $[f*g]' = f'*g + f*g' =$
$$(3x^2 + 10x - 3) * (7x^4 + x^2 - 12) + (x^3 + 5x^2 - 3x + 10) * (28x^3 + 2x)$$

Example 2:

Differentiate $\sqrt{x}\,(x^2 - 1)$.

solution: Let $f(x) = \sqrt{x}$ and $g(x) = x^2 - 1$. Then:

$$[f*g]' = f'*g + f*g' = [\sqrt{x}]'(x^2-1) + \sqrt{x}\,[x^2-1]' =$$
$$\frac{1}{2}x^{-\frac{1}{2}}(x^2-1) + \sqrt{x}\,(2x).$$

Just as we can take two simple functions and multiply them to get a new function, we can also divide them. Again, the optimist would hope that finding the derivative of two simple functions could be accomplished by some simple and straightforward formula. This is not he case. The situation is much worse than the one for the product.

Theorem 4 (quotient theorem): $\left[\dfrac{n}{d}\right]' = \dfrac{d*n' - n*d'}{d^2}$

The unfortunate thing about this equation is that both subtraction and division are involved. Any misplacement of terms will yield the wrong result. I have found that to reduce the chance of mixing up the terms, the following device helps:

n - numerator
d - denominator
Now, "d" is before "n" in the alphabet, therefore begin with "plain" d and end with "plain" d^2. Plain, meaning: not the derivative.

Example 3:

Differentiate $\dfrac{3x+1}{2x-1}$

d = 2x-1 and n = 3x+1, therefore d' = 2 and n' = 3

Using our quotient theorem:

$$\frac{d}{dx}\frac{3x+1}{2x-1} = \frac{(2x-1)(3)-(3x+1)(2)}{(2x-1)^2}$$

Example 4:

Differentiate $\dfrac{3x}{\sqrt[5]{x}+20}$

$d = \sqrt[5]{x} + 20 = x^{\frac{1}{5}} + 20$ and $n = 3x$

$d' = \dfrac{1}{5}x^{\frac{-4}{5}}$ and $n' = 3$

so the derivative is

$$\frac{(\sqrt[5]{x}+20)(3)-(3x)(\frac{1}{5}x^{-\frac{4}{5}})}{(\sqrt[5]{x}+20)^2}$$

Exercise 3.6

Find the derivative in the following problems.

1. $(3x+4)^2$

2. $\dfrac{3x+1}{5x-3}$

3. $\dfrac{5x^4}{1+2x}$

4. $\dfrac{q^2-3q+1}{2q}$

5. $(3x)^{\frac{1}{3}}$

6. $(16x)^{-.5}$

7. $\dfrac{q^5-q+1}{(4q)^{.03}}$

8. $(x-x^2)*x^{\frac{1}{3}}+x^{\frac{1}{4}}$

Section 3.7 The Chain Rule

We are almost finished learning the fundamental differentiation theorems. There remains one more that we need at this time. You can see the need for it quickly if I ask you to differentiate $h(x) = (x^2+20)^{100}$. A moments reflection should convince you that you have only one way to work this problem - expand it out. There has to be a better way! The secret lies in viewing the expression $h(x)=(x^2+20)^{100}$ not as one function but as two:

1. $f(u) = (u)^{100}$, and
2. $u(x) = x^2+20$

so $h(x) = f(u(x))$.

When the function is viewed in this fashion you can see that you have two functions **each of which** has the elementary form x^n, so can be differentiated by using the "elementary derivative number 1" rule. Hence,

$$\frac{d}{du}(u)^{100} = 100u^{99}$$

and

$$\frac{d}{dx}(x^2+20) = 2x$$

The problem is that we have now found the answers to two simple questions, $\frac{d}{du}(u)^{100}$ and $\frac{d}{dx}(x^2+20)$, and wish to know how these answers are related to the original question: find the derivative $h'(x) = (x^2+20)^{100}$. The answer is found in the chain rule theorem.

Theorem 5 (chain rule theorem): If $h(x) = f(u(x))$, then

$$\frac{d}{dx}f(x) = \left[\frac{d}{du}f(u)\right]\left[\frac{d}{dx}u(x)\right]$$

If you can express a function $h(x)$ as the composition of two or more functions $h(x)=f(u(x))$, then the derivative $h'(x)$ can be found by multiplying the derivatives of the component functions

$$\left[\frac{d}{du}f(u)\right]\left[\frac{d}{dx}u(x)\right].$$

After a little practice at decomposing functions you will see that the steps of this theorem can usually be carried out in your head.

Example 1:

Differentiate $(x^2+20)^{100}$

solution: Since we have already found that

1. $f(u) = (u)^{100}$,
2. $u(x) = x^2+2$,

and calculated $\frac{d}{du}(u)^{100} = 100u^{99}$ and $\frac{d}{dx}(x^2+20) = 2x$, the answer is $(100u^{99})(2x)$.

To get the final answer we need only substitute the definition of "u" back into this result. Hence, putting x^2+20 in for u we get

$$100(x^2+20)^{99}(2x).$$

Example 2:

Differentiate $\left[\dfrac{2x+1}{3x+8}\right]^{30}$.

First you should note that this will be a chain rule problem because you have a function raised to the 30th power. Thus:

1. $f(u) = u^{30}$

However, since $u(x)$ is a quotient:

2. $u(x) = \dfrac{2x+1}{3x+8}$

When finding the derivative of $u(x)$ you will need to invoke the quotient theorem. This situation is not at all unusual as you will see. It is often the case that two (or more) theorems will need to be invoked in a single problem.

solution: $\dfrac{d}{du}u^{30} = 30u^{29}$ and using the quotient rule we get

$\dfrac{d}{dx}\left[\dfrac{2x+1}{3x+8}\right] = \dfrac{(3x+8)(2)-(2x+1)(3)}{(3x+8)^2}$. Hence, the answer is

$$30u^{29}\left[\frac{(3x+8)(2)-(2x+1)(3)}{(3x+8)^2}\right].$$

Substituting the value of u back in we get:

$$30\left[\frac{2x+1}{3x+8}\right]^{29}\left[\frac{(3x+8)(2)-(2x+1)(3)}{(3x+8)^2}\right]$$

Are you ready for an observation which will make this process much quicker? First, recall something we stated some time ago:

When differentiating there are no situations where you take more than one derivative at a time.

Given that, let us differentiate the function $(x^2+20)^{100}$ again. If you are practiced enough by now to be able to view the component parts of the function, the result can be found as follows:

Differentiate $(\ )^{100}$ and multiply by the derivative of what is inside: (x^2+20). Since the derivative of $(\ )^{100}$ is $100(\ )^{99}$ and the derivative of x^2+20 is $2x$, the answer is:

$$100(\ )^{99}*(2x)$$

Notice that I have left what is inside the parenthesis blank. Since you must not calculate two derivatives at once, you will find it safest to only look at one function at a time. Viewing the outermost function as $(\ )^{100}$ will help you do this and hopefully you will avoid the error of taking two derivatives at once:

error $100(2x)^{99}$ **error**

Example 3:

Write down the derivative of $\sqrt[5]{3x^2+1}$ by the shortcut.

solution: Viewing this as $(\ )^{1/5}$ we get:

$$\frac{1}{5}(\ )^{-\frac{4}{5}}*(6x) \text{ or } \frac{1}{5}(3x^2+1)^{-\frac{4}{5}}*(6x)$$

Example 4:

Differentiate $(3x^2+2x+1)^5 * \sqrt{2x-9}$.

This problem requires both the use of the product theorem, for it is a product, and, as we differentiate each factor, the use of the chain rule. View the problem like this:

$$f(x) = (3x^2+2x+1)^5, \text{ and } g(x) = \sqrt{2x-9}.$$

In working the problem, we first invoke the product theorem to recognize that the result will be f'g+fg'. In the first term of this product, we have to find f' by using the chain rule. In the second term of the product we have to find g' by using the chain rule. Thus the final result will be:

$$f'g+fg' = \{[5(3x^2+2x+1)^4 * (6x+2)] * \sqrt{2x-9}\} + \{(3x^2+2x+1)^5 * \frac{1}{2}(2x-9)^{\frac{1}{2}} * 2\}.$$

Terminology needed in exercise 3.7

In some of these exercises you will need to use both the product rule and the quotient rule for finding derivatives while evaluating the derivative with the chain rule.

1. Product rule: $[f * g]' = f' * g + f * g'$
2. Quotient rule: $\left[\dfrac{n}{d}\right]' = \dfrac{d * n' - n * d'}{d^2}$

Exercise 3.7 Find the derivative of the following:

1. $f(x) = (x^2+1)^{50}$

2. $\sqrt{3x+1}$

3. $\sqrt[3]{5x^2 - 2x + 1}$

4. $g(x) = 3x(2x+1)^{10}$

5. $(2x+1)^5 * (3x-2)^7$

6. $\dfrac{2x+1}{(3x-2)^8}$

7. $\dfrac{(5x+2)^7}{2x}$

8. $\dfrac{\sqrt[5]{5x-1}}{\sqrt[3]{2x+1}}$

9. $\sqrt{5x-4} * (x^2-3x)^{10}$

Section 3.8 Using The Derivative To Find Marginal Values And Peaks And Valleys Of A Function

Let's start by comparing marginal cost and the derivative.

Example 1:

First, consider the following cost function:
$$C(q) = \frac{1}{75}q^2 - \frac{7}{6}q + 100$$

The cost table for the 101st item to the 105th item is:

q	101	102	103	104	105
cost	$118.18	$119.72	$121.29	$122.88	$124.50

Hence, the marginal cost of the 102nd item is:

$$C(102) - C(101) = \$119.72 - \$118.18 = \$1.54$$

The derivative of this cost function at q = 101 is very close to the marginal cost at q = 102. How close?

$C'(q) = \frac{2}{75}q - \frac{7}{6}$; therefore, C'(101) = $1.53. C'(101) and the marginal cost of the 102nd item are very close indeed.

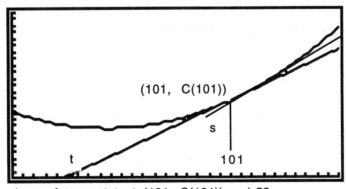

slope of tangent t at (101, C(101)) = 1.53
derivative of C(q) at (101, C(101)) = 1.53
slope of secant s passing through (101, (C(101)) and (102, C(102)) = 1.5
marginal cost of q = 102 is 1.54

Interpreting these results correctly we would say:

The marginal cost of producing **one more item** at the level q=101, or the **marginal cost of the 102nd** item is $1.54.

The marginal cost **at** the level of production q=101 is C'(101)=$1.53.

This leads us to a point of potential confusion and you must watch the wording very closely:

Because the marginal cost **of** the (q+1)st item (C(q+1) − C(q)) is so close to the marginal cost **at** the qth item (C'(q)), The derivative of the cost (or profit or revenue) is usually referred to as **the** "marginal cost" (or "marginal profit" or "marginal revenue") function. This marginal cost function evaluated at a point "q" will yield a value very close to the value of the marginal cost of the q+1st item. With this understanding most people will not make any careful distinction but will expect the audience to interpret the term "marginal ---------" correctly within the current context.

We now have one application of the derivative:

The derivative with respect to quantity (or slope function) can be used as a quick approximation for the marginal cost (or revenue or profit).

Another equally important application of the derivative is one that we have touched upon several times. At a peak or valley of a function the marginal value (or derivative or slope function) takes on the value 0. To put this in standard terms, peaks and valleys of a function correspond to **roots** of the derivative.

Example 2: Where do roots of derivatives occur?

Consider the function f(x)=x^2+1.

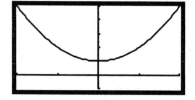

Note how the minimum occurs at exactly x=0. Without considering what kind of a graph the derivative should have, where should a root of the derivative occur? Since the slope of this function is 0 at precisely 0, you should realize that the derivative will have to have a root there.

Look at the straight line in the following graph. Finding the slope function to be f '(x)=2x, we have graphed both the function and the slope function on the same axis. Note from the graph how the root of f '(x)=2x has a root at 0 exactly where the minimum value of f(x)=x²+1 occurs.

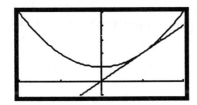

We could have done this problem fairly quickly by hand. We are interested in looking at more difficult problems and using our calculators to make the work easier.

Example 3: Finding peaks of the profit function

Let us now use the calculus on a profit function. Let a revenue function be defined by R(q) = 2q and a cost function defined by $C(q) = \frac{1}{75}q^2 - \frac{7}{6}q + 100$. Profit=Revenue-Cost; therefore,

$$P(q) = 2q - \left[\frac{1}{75}q^2 - \frac{7}{6}q + 100\right]$$

Since the production bounds where we can actually make a profit are so limited, we need to stay within those bounds. It is clear that we could find the end points or bounds of our profitable region by finding the break-even points (there will be two of them). These would be found by setting R=C or R-C=0 and solving.

However, an even more desirable point we wish to know is the peak of the profit function. To find this peak is to also find that production level q which maximizes profit. How do we find the q value for this point?

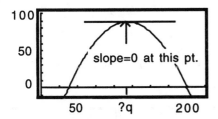

Observe that at the peak of the profit function, the slope of the curve is horizontal which yields a zero value. The slope is the same as the derivative. Thus, if we could find the value of q which makes the derivative zero we will have found the q value that makes the profit a maximum.

This is an easy problem to solve with the tools at your disposal. You are simply looking for the roots of P'(q) = 0. In this case, we hardly need a calculator. Since P'(q) = $2 - \frac{2}{75}q + \frac{7}{6} = \frac{19}{6} - \frac{2}{75}q$, therefore q = 118.75 is the solution. To check this out, Let's calculate the value of P(q) at three points; P(118), P(118.75), and P(119). What should we get for results? If 118.75 is the point of maximum profit, then both the values P(118) and P(119) should be smaller than P(118.75). Using EVAL in the GRAPH editor we get the following:

P(118) = 88.0133
P(118.75) = 88.020833
P(119) = 88.0200

Looking at the graph of the profit function and using the FMAX feature of the MATH menu (GRAPH, MATH, MORE, FMAX) we can see the following:

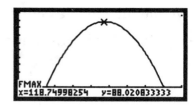

Notice that if we look at the ROOT of the graph of the derivative of the profit function (MP(q) = P'(q) = $\frac{19}{6} - \frac{2}{75}q$), we get the same quantity.

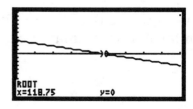

So, as advertised, finding the point where the slope of the profit function is 0 yields a maximum profit of 88.0217 at q = 118.75.

The root of the derivative or slope function can be used to find places where the original function is horizontal. This includes peaks and valleys.

How dangerous our algebra could be without a graph for you to find **both** peaks and valleys by this technique. Suppose our profit function had only a valley. We might have accidentally found the minimum profit since the slope is 0 there also! Care is needed and graphs will usually provide the additional information you need.

Example 4: Finding peaks and valleys in general

We are not limited to problems that yield linear or quadratic results. Consider the function:

$$h(x) = 4x^3 + 5x^2 - 3x + 1.$$

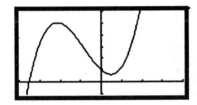

As you can see, this function has a distinct peak and a valley. Some questions we could ask are,

a. EXACTLY where do these peaks and valleys occur? That is, find the coordinates (x,h(x)) where these points are.

b. Do there exist other peaks or valleys elsewhere that we cannot see because of the limited nature of a graph?

a. We find that the derivative of $h(x) = 4x^3 + 5x^2 - 3x + 1$ is $h'(x) = 12x^2 + 10x - 3$.

Graphing the function and the derivative we have the following:

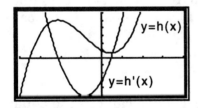

Note in the graph how the function h'(x) passes through the x-axis **DIRECTLY UNDER** the peak and valley. Thus, the roots of this derivative or slope function give the precise x-coordinate location of those points on the original function h(x). We can find the ROOTS of the derivative or use the FMAX and FMIN on the original function to find the x-coordinates of the peak (Max.) and the valley (Min.).

Either method will yield the same coordinates.

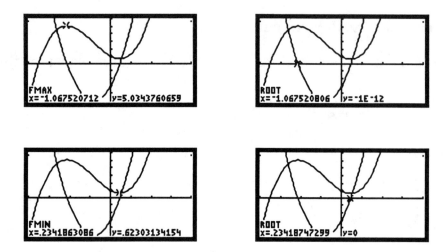

When we use the ROOT of the derivative method, we need to find the vertical component (the h value). To do this we substitute into h(x) and find

h(-1.06752)=5.0347 and h(.234187)=.623031

Thus, the peak shown in the graph for h(x) occurs at (-1.06752, 5.0347) and the valley at (.234187, .62303).

We can now answer the following important question we asked earlier:

b. Could there be other peaks and valleys for this function that are off the screen in our graph? The answer is no. Why? Because if there were any other peaks and valleys, the derivative would be 0 there. The derivative is a quadratic equation. Quadratic equations, being degree 2 polynomials, have at most two roots and we have found them both. Thus, the original function h(x) = 4x^3 + 5x^2 - 3x + 1 cannot have any peaks and valleys other than those that appear in our graph.

Example 5:

In a model for the annual expenditure on rental property one can use the equation $E=\frac{P}{x}+Rx^c$, where P is replacement cost, R is the average repair cost for the first year, and c is a positive constant reflecting increased repair cost as years, x, pass.

Consider the case of the laundromat. Given that a replacement machine costs about p = $500, and yearly repairs run about R = $30 and c has been calculated to be c =

4/3, how many years, x, should you keep the equipment to minimize E, the annual expenditures? You should first graph the equation $E = \frac{500}{x} + 30x^{\frac{4}{3}}$. What are reasonable scaling choices for x and E? We are talking about commercial washing machines, so some choice of x between 1 and 10 would be appropriate. Given the constant hard use in a laundromat it would seem appropriate to choose a number closer to 1. We will choose 2. Calculating the value of E at x = 2 we get about $325.60. Thus, a good scaling choice will include x = 2 and E = 325.60. We can use the following RANGE window to get the accompanying graph. Using the FMIN feature we find the optimum number of years to be 2.95191 with a minimum cost at that time of $296.42.

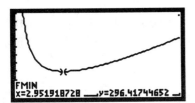 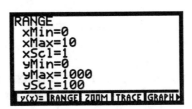

Finding the derivative $E'(x) = \frac{-500}{x^2} + 40x^{\frac{1}{3}}$, doing a graph, and finding the root, should give the same answer.

Information needed to work exercise 3.8

1. An understanding that the derivative of a function gives the slope of the function at any point.
2. Understanding that the roots of the derivative reveal horizontal spots on the function.

Exercise 3.8

Using the examples provided above to guide you, graph the functions in 1 - 5. Find the derivative and graph it on the same set of axes. Note the approximate location of the peaks and valleys and the corresponding roots of the derivative. Then find both the horizontal and vertical coordinates of the peaks and valleys by finding the root(s) of the derivative.

1. $f(x) = \dfrac{1}{3}x^3 - \dfrac{9}{2}x^2 + 18x$

2. $g(x) = \dfrac{x^3}{3} - x^2 - 3x$

3. Suppose the amount of a certain medication in the bloodstream t minutes after an injection is given by A(t) = 1500t / (t²+4). A patient is injected at 3:15 pm.

    (a) At what time is the drug's concentration a maximum?
    (b) How much of the drug is in the patient's bloodstream at 4:00pm?
    (c) At what time is the amount of drug in the patient's bloodstream less than 10 units?

4. $f(x) = \dfrac{x}{3}(x^2 - 75)$

5. $g(x) = \dfrac{x^3}{3} - \dfrac{3x^2}{2} - 18x + 40$, (Note that g'(x) is not affected by the addition of the constant; your graph just moves up 40 units)

6. Functions such as $C(q) = 25000 + \dfrac{450}{q} * 20 + \dfrac{3}{2}q$, represent the cost of ordering and storing goods (Economic Order Quantity curve). In this function, one needs to order 450 items over the course of the year but this may be done by placing several smaller orders of size q. If you have storage costs, larger orders will incur larger storage costs and in this function these costs are represented by the term $\dfrac{3}{2}q$. C(q) is the total cost for the 450 items incurred by placing (possibly many) orders of size q.

    a. What size orders q will minimize your cost, and
    b. What will be the cost there?
    c. Convince yourself that you have found the minimum cost by calculating C(q) for two other values of q; q = 450 (place one BIG order), and q = 50 (more and smaller orders).

7. A demand function is given by p = –10q+100.

   a. Find the maximum revenue.
   b. Compare the marginal revenue of the 6th item with the marginal revenue at the 5th item found by using the derivative.
   c. Give some explanation of these results.

8. A cost function is given by C(q) = 5q+100 and a demand equation by p = –8q+300. Find the profit peak.

9. In problem 8, compare the marginal profit **of** the 10th item with the marginal profit **at** the 9th item.

10. Fit a cubic equation to the following set of data and find the peak(s) and valleys:

horiz.	20	30	40	50	60
vertical	1.3	3.8	-0.1	1	7

Graph the function in 11 and 12; then find any peaks and/or valleys.

11. $f(x) = \dfrac{1}{1-x^2}$

12. $f(x) = \dfrac{x^3 - 4x^2 + x + 6}{x^2 - 5x + 6}$ Could this one be a trick question?

13. The following sales (quantities) of turkeys were observed during an 8 month period. Fit a polynomial of degree 4 to this data.

month	5	6	7	8	9	10	11	12
sales	20	23	12	7	13	35	70	60

   a. Graph showing the points and the function.
   b. On a scale of 1-10 (1 bad, 10 good), how do you like the fit?
   c. Give the value of the peaks and/or valleys of your function in the interval in question.
   d. According to your function, during what month(s) did the maximum sales occur? How does this correspond to the actual data?

Graph the functions in 14 - 19 and find any peaks and/or valleys.

14. $f(x) = \dfrac{(x^2(9-x)^3)}{270}$

15. $g(x) = -(80-x)^3 \left[\dfrac{6x}{500}\right]$

16. $f(x) = \dfrac{x^5}{2} + 3x^4 - \dfrac{2(x+3)^2}{3} - 7x - \dfrac{79}{50}$

17. $(2x^2 - 6x)^{\frac{1}{3}}$

18. $g(x) = -2x^4 + 9x^3 + 2x^2 - 6x + 31/20$

19. Given the function $f(x) = x^4/3 + x^3/5 - 5x^2 + 4x - 6$.

20. Suppose you have a rental car agency and wish to minimize the value $E(x) = \dfrac{P}{x} + Rx^a$, where E is the annual expenditures for repairs and replacement, x is the average life of the product in years, P is the replacement cost, R is the average repair bill for the first year and "a" reflects increasing repair costs with the passage of time. Assume that the correct model is:

$E = \dfrac{17000}{x} + 400x^{\frac{4}{3}}$. Graph E, E' and label your graph indicating how many years should you keep this car to minimize E.

21. If the mathematical model for annual expenditures on your laundromat washers is given by $E(x) = P/x + Rx^a$, where E is the annual expenditures for repairs and replacement, x is the average life of the product in years, P is the replacement cost, R is the average repair bill for the first year and "a" reflects increasing repair costs with the passage of time, how many years should the laundromat keep its machines to minimize E? For your new washing machines E is about $E = \dfrac{500}{x} + 10x^{\frac{5}{4}}$. Graph E, E', and find the optimal number of years before replacement to minimize overall expenditures.

22. ASUTravel offers group rates for a trip to New York as follows: The minimum group size is 200, the maximum is 300. If there are 200 in the group, the fare is $400. For every person beyond 200 they will cut the fare by $1 per person. Their cost is $75 per person.

a. Find the revenue, cost and profit functions, then graph the curve.
b. What size group will maximize their profit?
c. What will each person's ticket cost?

23. ASUTravel offers group rates for a trip to Moscow as follows: The minimum group size is 200, the maximum is 300. If there are 200 in the group, the fare is $450. For every person beyond 200 they will cut the fare by $1.30 per person. Their cost is $200 per person.

a. Find the revenue, cost and profit functions, then graph the curve.
b. What size group will maximize their profit?
c. What will each person's ticket cost?

Section 3.9 Maximums And Minimums With Constraints

Given a profit function we will normally be interested in the maximum within some interval. Likewise, given a cost function we will be interested in finding the minimum within a certain interval. We will frequently need to find the slope function to find candidates for maximums and minimums that occur.

In finding the peaks and valleys we must also consider the fact that most business situations have real world constraints. Some examples would be:

Time constraints: there are only 24 hours in a day. Hence, if you are doing something on a daily basis, only 24 hours can be used.

Financial constraints: perhaps the best thing is to buy 100 machines and run them all the time, but if you can only get enough money for 20, then that is the best you can do.

Machine constraints: While it is theoretically possible to run a machine 24 hours a day, 7 days a week, lack of down time for maintenance will eventually prove to be very expensive. For example, don't stop to change the oil, burn up the engine.

Personnel constraints: Only an idealist would think humans can work at a dull job efficiently for an indefinite period of time.

Each of these constraints will impose an interval within which we must work. All solutions we find must fall within these intervals. For example, consider the following graph. It is the graph of a function with the constraint o≤x≤b.

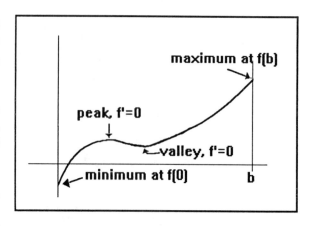

In this example, the minimum occurs at the left end point, zero, and the maximum occurs at the right end point, b. This is where graphing really pays off. Your colleagues doing this by hand will have to laboriously calculate all the values, including those for the peaks and valleys, to see what will yield the largest and smallest functional values in the interval [0,b]. You will graph the function and simply calculate f(0) for the minimum and f(b) for the maximum. There is simply no point in doing any calculus since you can see that the values we are interested in do not occur at the peaks and valleys. With this warning about checking your graph before proceeding, let us proceed to solve a few problems that do involve calculus.

We have worked a little with the concepts of revenue, cost, and profit. We now need to see how these concepts fit in with the ideas of calculus we have developed to date. As you know, Profit=Revenue-Cost. In many business situations we wish to minimize cost and maximize revenue. A constant problem is that minimizing your cost can have a deleterious effect on your revenue; so much so that any savings accrued in the cost function are more than offset by the loss of revenue. The solution to this situation is to look at the profit function. Since the profit function takes into account both revenue and cost, maximizing the profit function will automatically balance the revenue versus cost question.

Using standard business terms, the slope function or derivative of a function is called the "marginal _____". Thus, the derivative of the profit function is called the **marginal profit**, the derivative of the revenue function is called the **marginal revenue**, etc. Until now we used the term "variable cost" when talking about the slope of a linear cost function.

Consider the linear cost function 3x+70. "3" would be the variable cost, but since it is the slope of the equation it would also be the "marginal cost". In general, you should not think of variable coast and marginal cost as the same. They will be the same only when we are considering equations which are polynomials.

Example 1: Maximum profit with constraints

For review purposes, suppose we have cost and revenue functions defined as follows:

Cost: $C(q) = 7 + 5q + .1q^2$.
Revenue: $R(q) = 10q - .1q^2$.

both defined on $0 \leq q \leq 20$

Our profit function would be

$P(q) = R(q) - C(q) = 10q - .1q^2 - (7+5q+.1q^2)$.

Reducing the equation and graphing (xMin=0, xMax=20, xScl=1, and ZFIT for y window) we get the accompanying graph.

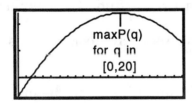
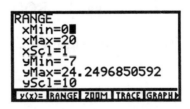

With our knowledge of calculus we can see that the best profit will occur at the peak of this function. This peak has a unique characteristic that we can use calculus to find. Remember that the derivative gives the slope of the equation at any point. In this equation, we note that the maximum profit occurs at the peak and the slope there is 0.

If we find the marginal profit, set it equal to 0 and solve for q, ie. find that root of the derivative, we will have the q coordinate where profit is a maximum. The P coordinate can be found by substituting in the q value.

Calculating the derivative yields P '(q) = -.4q + 5. Setting this equal to zero and solving gives q = 12.5. P(12.5) = 24.25. Thus, a maximum profit of $24.25 is achieved at a quantity of 12.5 as shown in the graph.

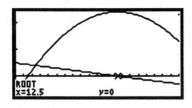

which gives us the same if we use the FMAX feature on the profit function.

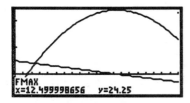

Looking at the first graph, observe how we have found the peak of this curve by finding the **root** of the slope equation, P'(q). This interchange between the equation and its derivative is often a point of confusion for students. Observe carefully how we set the **derivative** P' equal to 0 and solved for q to get q=12.5. This gives us the q value of the peak of the profit function, but **not the profit there**. In order to find the profit, we have to put the q value 12.5 back into **the original equation,** P(12.5) to get the maximum profit, $24.25. In outline form, the first procedure we have used is as follows:

1. Differentiate the function P(q)
2. Set the derivative P'(q) = 0 and solve, or
   graph the derivative and use the ROOT feature to find the q where the derivative = 0.
3. Take this root r of P'(q) and calculate P(r). The value we are looking for is (r, P(r)).

Using the second method (FMAX feature) we obtain the quantity q and the maximum profit P(q).

Now suppose we had constraints of 0≤q≤10. How would this change the problem?

If your graph is accurate enough, you can spot that the maximum occurs **at the right end point**. Since this is the case, we would calculate P(10) and stop, since there is no need for any calculus.

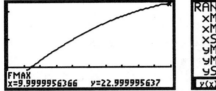 

Example 2:

Suppose a firm has a cost function $C(q) = 2500 + 25q + q^2/2$ and a revenue function $R(q) = (100 - q/2)q^2$. Find the break-even point(s) and the maximum profit in the interval $0 \leq q \leq 225$.

solution: A general outline of the solution would be as follows:

In the GRAPH editor:

a. Enter the revenue as y1.
b. Enter the cost as y2.
c. Enter the profit as y1-y2. Deselect the y1 and y2.
d. Set the RANGE window xMin = 0, xMax = 225, xScl = 25, ZFIT for y values.
e. Use the ROOT feature to find the break-even point(s).
f. Use the FMAX to find (q, P(q)) for maximum profit.

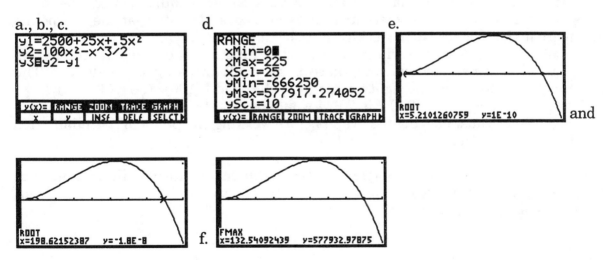

Additional consideration: ANOTHER INTERVAL RESTRICTION

Suppose that because of time (you can't run more than 24 hours per day) or other restrictions, a production limit of 100 units was imposed. If this were the case, where would the maximum profit occur?

Example 3: More on maximum profit with constraints

Suppose a firm has a cost function $C(q) = 2500 + 25q + q^2/2$ and a revenue function $R(q) = (100 - q/2)q^2$. Find the maximum profit in the interval $0 \leq q \leq 100$.

solution: If you were to solve this problem without benefit of a graph, you would need to find and evaluate all possible points where the maximum profit would occur.

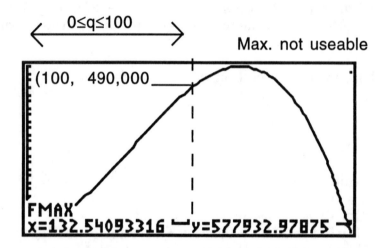

This could be done using the derivative However, note that the additional condition $0 \leq q \leq 100$ may seriously affect our possible answers. In fact, a quick look at the graph shows how seriously. This is where a graph is most useful. Note that the end points take on great significance when working a problem such as this.

To solve this problem we will first graph the function and "eyeball" the graph. Doing this it becomes obvious that there is no point in using any calculus. The maximum profit occurs at the right end point, $q=100$ and $P(100)=\$504,700$.

Hence, even though $P(132.5)=\$577,924$ would be a better solution, this solution is not a possibility. Thus, viewing the graph before working the problem has eliminated any need of using calculus.

An interesting class of problems that usually have interval limitations are **Economic order quantity** problems. As an introduction to this class of problems, consider the following example.

Example 4: Economic order quantity, batteries

You decide that over the course of the next year you can sell 50,000 flashlight batteries. Your marginal cost is 85¢ each. However, each order you place costs $40. In addition, your storage costs are 5¢ on the average inventory. Initially we will not place any size restrictions on orders (no interval considerations). How many orders of size "x" should you place to minimize your overall costs?

solution: To resolve this problem you must recognize that it consists of three separate cost functions:

1. Overall cost which is 50,000*0.85. No matter what your strategy is, this amount will eventually have to be paid.

2. Order cost or batch cost (depending upon the situation, sometimes called "set-up costs"). Since each order you place costs $40, you must calculate how many orders you are going to place and multiply that times $40: (# of orders)*(cost of each order). To calculate the number of orders, consider how many orders you would place if "x" were 50,000. Clearly you would place 1 order. If you placed orders of size 25,000 you would place 2 orders. You can see that the number of orders you place is calculated by using the fraction $\left(\dfrac{50000}{x}\right)$, hence, this part of your cost can be calculated by $\left(\dfrac{50000}{x}\right)*\$40$.

3. Storage costs. To calculate these costs you have to understand the term "average inventory". Average inventory refers to how many items you have in stock on average. For example, if you order x items and start selling them off, at the end of some time period you have none left. Over this time period, what was the average number you had in stock? The answer is $\left(\dfrac{x}{2}\right)$.

To help you understand why this is, consider the following graphical representation of the situation.

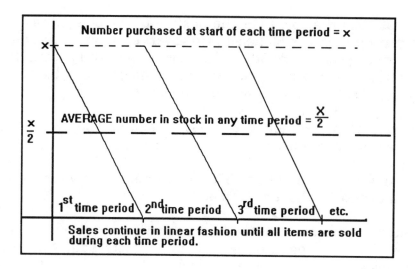

Hence, the cost for storage is $\left(\dfrac{x}{2}\right)*.05$.

To calculate the final cost, C(x) of placing several orders of size x, we must add each of the above costs:

$$C(x)=(\text{overall cost})+(\text{batch cost})+(\text{storage cost})$$

This yields the equation:

$$C(x) = 50{,}000 * 0.85 + \left(\dfrac{50000}{x}\right)*\$40 + \left(\dfrac{x}{2}\right)*.05$$

Problems such as this are often referred to as **"inventory problems"** since the unwary merchant will jump at the chance of making a larger profit by volume purchasing without realizing that it costs money to have stock in inventory and a large inventory can destroy you. Once you have come up with the correct cost equation, graph, and use FMIN to find x (the number in each order). A logical RANGE window: xMIN=0, xMAX=50000, xScl=10000, then ZFIT.

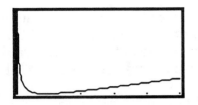

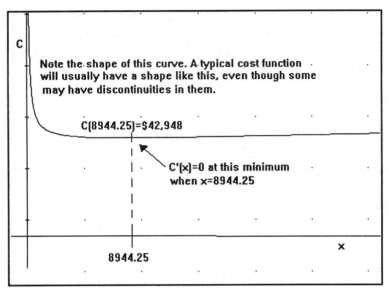

Note the shape of this curve. A typical cost function will usually have a shape like this, even though some may have discontinuities in them.

C(8944.25)=$42,948

C'(x)=0 at this minimum when x=8944.25

8944.25

To answer the question of how many orders, 50000/8944.25 = 5.59

Now consider the following small change in the problem: The merchant can't store more than 5000 batteries at one time due to space considerations. Exactly where the minimum occurs with respect to 5,000 can be difficult to spot in the graph. The "best" choice of 8944.25 is close enough to 5,000 that this difference can be difficult to spot in the graph. In this problem, the safest way to proceed is to do the substitution and check the results:

C(8944.25) = $42,947, and

C(5000) = $43,025

The difference in total cost in this case is small. You could easily conclude that worrying about additional storage to decrease cost is not worth the effort.

Finally, let us end this section with a little philosophy. As you know, we have been stressing the idea of finding the maximum profit, minimum cost, etc. In a capitalistic society this is very important. A good return on your investment guarantees money for future expansion and development thus providing more jobs and more profits.

Another consideration is the fact that one can often affect the number of jobs available by attempting to maximize profits, ie., under some circumstances we can increase jobs by increasing production and settling for a little less profit. This strategy would tend to push the profit more towards the second break-even point and reduce profit. Under what circumstances would this be desirable? If you had to make such a decision what would you do?

More than one businessman has had to make such a difficult decision. It often boils down to having happy workers and unhappy stockholders or people out of work and happier stockholders. Greater efficiency with less workers, less efficiency (at least in the short run) and more workers. One can even internalize this into a clear conscience vs. a very troubled one. What would you do?

Information needed in exercise 3.9

1. Total variable cost = (variable cost)*quantity.
2. Revenue = (demand price)*quantity
3. Break even points occur where $R(q) = C(q)$. In solving on the calculator, find root of $P(q)$, since $P(q)=R(q)-C(q)=0$.
4. EOQ: $C(x)$ = (overall cost) + (batch cost) + (storage cost)

Exercise 3.9

1. Suppose you have found the demand function to be $p = -4q + 200$. In addition suppose we know that the cost function $C(q)$ has variable cost $= \sqrt{\dfrac{2000}{q}}$ per unit and fixed cost $500. Answer the following questions. Be careful when you enter $\sqrt{\dfrac{2000}{q}}$ into the calculator that you do not accidentally enter $\dfrac{\sqrt{2000}}{q}$.

   a. Find and graph the revenue function.
   b. Find and graph the cost equation.
   c. Observe in your graph about where the break-even points are.
   d. Find the profit function.
   e. Graph the profit equation and note where the roots are.
   f. After observing that the roots of the profit equation are in the same location as the break-even points ($R(q)=C(q)$ when $P(q)=R(q)-C(q)=0$), find the roots of the profit equation.
   g. Find the value of q which yields maximum revenue and the revenue there, and
   h. Find the value of q which yields maximum profit and the profit there.
   i. In general, will maximum revenue or maximum profit be greater?

2. In problem 1, assume that you may produce only between 0 and 20 units. Now:

   a. Find the value of q which yields maximum revenue, and
   b. Find the value of q which yields maximum profit.

3. Given a selling price of $p = 2000-20x-x^2$ and a cost function $C(x) = -100(x-10)^2+10000$, defined on the interval from 0 to 50,

   a. what would be the best level of production x, and
   b. the maximum profit there?
   c. If you could increase production to 100 items, what would be the best level of production and the profit there?

4. Given the demand curve for an item is $D(x) = p = \dfrac{10}{7x+1.5} + 9$, and the supply curve $S(x)$ is a linear function which passes through the points $(x_0, s_0) = (-1, 0)$ and $(x_1, s_1) = (4, 3)$.

   a. Draw a graph of both demand and supply curves. Observe where the equilibrium point(s) are.
   b. By finding the root(s) of $D(x)-S(x)$ and substituting these values into either $D(x)$ or $S(x)$, find the equilibrium point(s).

5. Given that a demand function $D(x) = \dfrac{11}{5x+.5} + 3$, we get a revenue function $R(x) = (\dfrac{11}{5x+.5} + 3)x$. Derive the cost equation using a variable cost of 2 per unit with a fixed cost of 20 and find the break-even point(s).

6. Let the demand price be $p = 8000 - x$ and the cost equation be determined by a variable cost of $4000+5x$ per unit with a fixed cost of 800. Answer the following questions.

   a. Find and graph the profit function $P(x)$.
   b. The value of x where $P(x)$ is a maximum and the profit there.
   c. The marginal profit **at** x = 200.
   d. The marginal profit **for** the 201st item.
   e. Using the graph and the numerical data you find, write a paragraph relating what you think the best strategy would be for production for the given profit function.

7. A company has a selling price of p = 150x-8x² with a cost function which has variable cost = 20 per unit and fixed cost = 30. The number of items produced usually varies from 0 to about 20.

   a. Find and graph the profit function
   b. Find the break-even point(s).
   c. Exactly how many items should this company produce to maximize their profit and what is the profit there?
   d. What is the marginal profit **at** x=10?
   e. What is the marginal profit **for** the 11th item?

8. Given that the demand price for an item is $D(x) = p = \dfrac{11}{5x+.5} + 3$, and the supply price S(x) is linear and passes through (0,0) and (5,2),

   a. Draw a graph.
   b. Find the equilibrium point.
   c. Find the revenue function

9. Given a demand function p = 250x-16x² and a cost function determined by **total variable cost** = 40x and fixed cost = 60 (so the cost equation will be C(x) = 40x+60).

   a. Find the break-even point.
   b. Exactly how many items should this company produce to maximize their profit, and what is the profit there?
   c. What is the marginal profit at x=5?
   d. What is the marginal profit for the 6th item?

   Explain what you do and draw a graph to justify your answers.

10. Consider a product for which the minimum factory direct purchase is an 18 wheeler full. A direct purchase yields a cost of 5¢ a linear foot, plus $1500 shipping per order, regardless of the number of linear feet in the truck (half a truckload costs the same to ship as a full one) plus storage costs of 7¢ per linear foot per year on the average inventory. On average you use about 200,000 linear feet per year.

    a. Write the equation for placing (and shipping) orders of any size. Draw a graph of this equation and calculate the size order that will yield minimum cost and what is that minimum cost.

b. Now suppose that a truck can carry at most 5000 linear feet. The answer to part a. yields a result larger than 5000 linear feet. If a truck can carry 5000 feet or less, what strategy should you follow, and what is the total cost of this strategy?

c. Suppose that any amount can be shipped for the cost of $1500 (no 5000 foot limitation). Hence it will be economical to place orders which minimize cost. However, the company has storage room for at most 10,000 linear feet ($0 \leq x \leq 10,000$) at a time. Now what is the best sized order? Give the exact difference in cost between this strategy and the optimal strategy found in part a..

d. Given this difference in cost, what might be a good strategy and why? What pitfalls might there be with this strategy?

11. Ice scrapers cost $1.20 each with a batch cost of $200 per order, regardless of size. The storage costs are 10% ($0.12) of the average inventory costs. If you predict sales to be 50,000 during the course of the next year,

   a. Write the cost equation for placing orders. Draw a graph of this equation and calculate the size order that will yield minimum cost and calculate that minimum cost.
   b. Given the additional restriction that storage considerations restrict order size to 5,000 or less, what size order is best? What is the cost for this size order?
   c. Should the merchant consider taking action to increase storage?

12. You have decided to start a company to manufacture fizzles. In checking out the fizzle market, you discover that the demand curve responds to market conditions as follows: At a price of $5.98 the demand is about 11,000. At a price of $9.98 the demand is about 8,000, at a price of $12.98 demand is about 6,000 and at a price of $8 the demand is about 9,000. In calculating your costs, you come up with a variable cost per unit of $4 and fixed costs of about $10,000.

   a. Graph the given demand points and make a best choice for a demand function. Then find that demand function.
   b. Find and graph the profit function.
   c. Find the break-even points.
   d. Where would your maximum profit occur and what is your profit there?
   e. Should you actually start up your company? Explain your reasoning.

13. In problem 12, you decide to investigate becoming a fizzle middle-man instead of a fizzle manufacturer, using the telephone to receive orders, and then placing lump orders with a supplier in Borneo. The supplier in Borneo will sell to you fizzles at a cost of $3.75 per item plus a flat $500 per order. You predict that you will have orders totalling about 4,000 items during the first year. Storage costs run about $2.50 on the average inventory.

a. Write down your total cost function in terms of order size.
b. Using the revenue function you found in example 12, calculate R(4000).
c. Using your total cost function, find the cost of placing one order of size 4,000.
d. Using calculus, find your optimal sized order and the cost of placing orders of that size.
e. Using the revenue at your predicted sales level of 4,000 find your best profit and make a comment about whether or not you should become a fizzle middle man.

14. You have decided to start a company to manufacture zizzles. In checking out the zizzle market, you discover that the demand curve responds to market conditions as follows: At a price of $5.98 the demand is about 12,000. At a price of $9.98 the demand is about 8,000, at a price of $12.98 demand is about 1,000 and at a price of $8 the demand is about 9,000. In calculating your costs, you come up with a variable cost per unit of $5 and fixed costs of about $30,000.

   a. Graph the given demand points and make a best choice for a demand function. Then find that demand function.
   b. Find and graph the profit function.
   c. Find the break-even points.
   d. Where would your maximum profit occur and what is your profit there?
   e. Should you actually start up your company? Explain your reasoning.

15. In problem 14, you decide to start up your company as a zizzle middle man, using the telephone to receive orders, and then placing lump orders with a supplier in Borneo. The supplier in Borneo will sell to you zizzles at a cost of $4.50 per item plus a flat $500 per order. You predict that you will have orders totalling about 4,000 items during the first year. Storage costs run about $2.50 on the average inventory.

   a. Write down your total cost function in terms of order size.
   b. Using the revenue function you found in example 14, calculate R(4000).
   c. Using your total cost function, find the cost of placing one order of size 4,000.
   d. Using calculus, find your optimal sized order and the cost of placing orders of that size.
   e. Using the revenue at your predicted sales level of 4,000 find your best profit and make a comment about whether or not you should become a zizzle middle man.

CHAPTER 4: EXPONENTIAL AND LOGARITHMIC FUNCTIONS

Section 4.1 Future Values and Present Values

In a prior section we learned how to differentiate one elementary function, x^n. Fortunately for you, there are only three really distinct types of functions which occur commonly in business. The first is the one we have done, x^n. The next two are the functions e^x and $\ln x$. You have no doubt already worked with these functions, but let's review for a moment what they mean and what their graph looks like.

"e" is an irrational number that has the value 2.71828•••, a never ending, non-repeating decimal. Mathematicians are interested in how it comes about, which we will look at briefly, but in business we are interested in the uses of this number. In business, it occurs when one tries to calculate compound interest. We start by calculating the compound interest formula.

Let F = future value, P = present value, r = interest rate in decimal form. Then if you invest (deposit) P dollars at r interest, at the end of year 0 (at the start of the first year) you have:

$F(0) = P$ and at the end of the first year you have:

$F(1) = P + rP = P(1 + r)$. At the end of the second year you have the amount of money you start the second year with, $P(1 + r)$ plus the interest on that amount which is $rP(1 + r)$ or:

$F(2) = P(1 + r) + rP(1 + r) = P(1 + r)^2$. At the end of the third year you would have the amount you start the year with $P(1 + r)^2$ plus the interest on that amount, $rP(1 + r)^2$ or:

$F(3) = P(1 + r)^2 + rP(1 + r)^2 = P(1 + r)^3$. Even without being a mathematical wizard you can see the pattern here and determine that after t years the future value will be:

$F(t) = P(1 + r)^t$. This is the formula for annual compound interest.

Back in the good old days before the advent of "creative banking" (translation: you, the public, are going to have a tax debt burden from lending institutions that made bad loans until you die) and banking deregulation, the amount of interest that could be paid by savings and loan institutions was fixed by law. For example, for many years you could receive only 5% interest (or less). In trying to be competitive for your money, a lending institution would try any gimmick they could to get your money.

I can still remember the first bank in our little town that thought of giving interest compounded semi-annually. My mother, being the family money manager, immediately transferred most of our funds to this bank. The formula for this compounding is

$$F(t) = P(1 + (r/2))^{2t}.$$

To give you some idea of the difference, suppose you had $10,000 in the bank for 30 years at 10% interest compounded yearly. This would yield $174,494.02, whereas the same money compounded semi-annually would yield $186,791.85, for an increase of $12,297.83. This dramatic difference was not lost on my mother.

Not to be outdone, the other bank in town (there were only 2) decided to offer interest compounded quarterly to try to regain some customers, and it worked. My mother immediately transferred most of our money back to the original bank. The formula for this is:

$$F(t) = P(1 + (r/4))^{4t}.$$

This will yield another increase:

ANNUALLY	SEMI-ANNUALLY	QUARTERLY
$174,494.02	$186,791.85	$193,581.49

for an additional increase of $6789.64.

The other bank soon followed suit with quarterly interest. I'm sure the end of the competition had something to do with the screams of pain from the bookkeepers. Things remained at this impasse until the advent of the high speed programmable digital calculating machine (I believe we now call them computers). With such power at their fingertips, the war of interest periods heated up again. Many institutions immediately went to monthly interest, some went to daily and discovered that could even cause a computer problems. Then some bank (does anybody know who?) was fortunate enough to have a programmer who was a mathematician and also knew some business. This fortunate institution was the first to offer **continuous** interest.

The term "continuous interest" means that interest is calculated on your principle every instant. From the work we have done, you can see that monthly interest would have the equation:
$F(t) = P(1 + (r/12))^{12t}$, daily interest the equation: $F(t) = P(1 + (r/365))^{365t}$, so continuous interest would be $F(t) = \lim_{n \to \infty} P(1 + (r/n))^{nt}$. As you can see, evaluating this limit (which depends only on n) boils down to trying to figure out how to evaluate the limit: $\lim_{n \to \infty} (1 + (r/n))^{n}$. This is where a deeper knowledge of mathematics is required. It has been known by mathematicians for several hundred years that this limit is e^r. Putting the P and t back into our equation, we see that the formula for continuous interest is: $F(t) = Pe^{rt}$. Going back to our example of $10,000 invested for 30 years at 10% = .1 interest, and calculating the value of $F(30) = 10000e^{(.1)30}$, we get the

following comparison:

ANNUALLY	SEMI-ANNUALLY	QUARTERLY	CONTINUOUS
$174,494.02	$186,791.85	$193,581.49	$200,855.36

An immediate consequence of continuous interest is that the interest received is greater. A not-so-obvious effect is that since the interest is calculated more frequently, businesses whose account tends to fluctuate widely will receive more interest on their money. It is typical of lending institutions to want to calculate interest on the minimum balance during a particular time frame. I still recall the shock I got when I received interest from an institution that gave monthly interest on the (**very** small print here) minimum balance for the month. I had kept about $20,000 in the account for 29 days, and had $700 for the last two. My interest was on the $700. I was not a happy camper. Within two days I was no longer their customer.

We have now derived the following collection of equations:

INTEREST ON "P" DOLLARS COMPOUNDED at RATE "r", "n" TIMES PER YEAR FOR "t" YEARS:

$F(t) = P(1 + (r/n))^{nt}$ (We will store this formula in our calculator as **F1**.)

INTEREST ON "P" DOLLARS COMPOUNDED CONTINUOUSLY AT RATE "r" FOR "t" YEARS:

$C(t) = Pe^{rt}$ (We will store this formula in our calculator as **F2**.)

Example 1: Finding future values (TI-82 users see Calculator Instructions III Section 4.1 for TI-82)

Find the future value of $1,000 at 6% interest for 20 years compounded:

    a. annually   d. monthly
    b. semiannually   e. daily
    c. quarterly   f. continuously

You can work these problems by simply substituting the various quantities into the correct equation and then using your calculator to compute the answer. However, when you have a series of problems to do, you should store the formulas in the SOLVER of the calculator. To do this follow these steps:

1. On the HOME screen, enter the formulas and press ENTER after each is entered.

   F1=P(1+(r / N))^(N*t) , ENTER ("n" is not an acceptable name for a variable, therefore use "N") We should see "Done" when this is completed.

   F2=P*e^(r*t) , ENTER

2. Press SOLVER (2nd GRAPH). We should see "F1" and "F2" located over the menu keys.

3. Clear any existing function or equation beside "exp". Press RCL (2nd STO), F1, then ENTER, ENTER. Clear any unwanted values that are beside the variables. We should see the following:

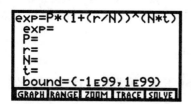

4. Now, simply substitute in for the known variables, place cursor beside the unknown variable and press SOLVE.

Hence, the future values of $1,000 at 6% interest for 20 years compounded annually, semiannually, quarterly, monthly, and daily is:

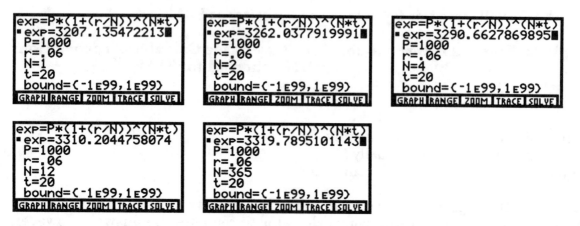

For continuously, move the cursor up to the top line and press CLEAR, RCL, F2, ENTER, ENTER.

```
exp=P*e^(r*t)
■exp=3320.1169227365■
 P=1000
 r=.06
 t=20
 bound=(-1E99,1E99)
■left-rt=0
GRAPH RANGE ZOOM TRACE SOLVE
```

PRESENT VALUE:

It is often the case that we want to know the **present value** of some future amount of money or property. For example, suppose you wish to know how much money you would need to put in the bank now at 8% interest to have $7,000 5 years from now? The answer to this and similar questions can be found by solving the equations for the future value for P.

a. Given the interest is annual, if $F=P(1+r)^t$, then solving for P we get $P = \dfrac{F}{(1+r)^t}$

or $P = F(1+r)^{-t}$.

Note that in the final result the only difference between the two equations is the negative sign in the exponent. This is sometimes a point of confusion so you need to make a note of this fact now since this will occur in each case.

b. Given the interest is quarterly, $F1 = P\left(1+\dfrac{r}{4}\right)^{4t}$, so $P = F1\left(1+\dfrac{r}{4}\right)^{-4t}$

If interest is compounded "n" times per year, the function definition would be:

Present value of F2 dollars at interest r compounded n times per year discounted back t years:

$$P = F1\left(1+\dfrac{r}{n}\right)^{-nt}$$

Finally, we have the case for continuous interest:

c. Given the interest is compounded continuously, $F2 = Pe^{rt}$, so $P = F2 e^{-rt}$.
   For interest compounded continuously the definition would be:

Present value of F2 dollars under continuous interest at rate r discounted back t years:

$$P = F2 e^{-rt}$$

Depending upon who is talking about the present value of something, different terminology may be used. For example, you will frequently find the term "discounted back" is used. A common question is:

"What is the value of $6,000 discounted back 10 years assuming a rate of 9%?"

Example 2: Finding present values (TI-82 users see Calculator Instructions III Section 4.1 for TI-82)

Find the present value of $6,000 at 9% interest for 10 years compounded:

  a. annually
  b. semiannually
  c. quarterly
  d. monthly
  e. daily
  f. continuously

To solve these problems in the SOLVER of the calculator we need only assign values to the known variables and solve for "P" instead of "F1" or "F2". Hence, the present value of $6,000 at 9% interest for 10 years when interest is compounded annually, semiannually, quarterly, monthly, daily and continuously is:

Terminology needed in exercise 4.1

1. Future value equations:
   F1=P(1+(r/N))^(N*t), F2=Pe^(r*t)

2. Present value equations:
   P=F1(1+(r/N))^(-N*t), P=F2e^(-r*t) Note: We do not need to enter these formulas in our calculator. We will solve for P in the formulas F1 and F2 in the SOLVER of the calculator.

Exercises 4.1

1. Assuming the stock market pays an annual long-term return of 11% annually. How much money would you need to invest today to have $2,000,000 at your age 70? (Assume your age is 20.)

2. The grandparents would like to help with the expenses of a new granddaughter's college education. How much will they need to invest now at 7.25% compounded quarterly so that they will have $30,000.00 in eighteen years?

3. Find the future value of $1000 invested at 8% interest for 20 years compounded

    a. yearly
    b. quarterly
    c. continuously

4. Find the future value of $12,000 invested at 6.75% interest for 30 years compounded

    a. yearly
    b. quarterly
    c. monthly
    d. continuously

5. Find the value of $100,000 discounted back 20 years at a continuous rate of 7.5%.

6. How long will it take an investment of $8500 to mature to $25,000 at a rate of 7% compounded

    a. yearly
    b. quarterly
    c. monthly
    d. continuously

7. Laws have been passed stating that after an account in a bank is inactive for a certain period of time, the bank should make every effort to find the owner; but, if unable to this in some fixed time the money in the account reverts to the state. These laws were motivated by several situations, one being a bank in an older city finding that a certain John Doe had deposited $750 in the bank in 1678. The bank was legally bound to find the heirs and distribute the money plus interest to them. During the first 280 years, interest was fixed at 4.5%; for the next 10 years the interest was 7.5%, at which point the bank calculated how much money would need to be distributed. How much was this amount? Do

you see why states have such laws?

8. A law is under consideration now that would allow money set aside for the educational purposes of a particular child to be tax free upon withdrawal. Assuming a combined state and federal tax rate of 38%,

   a. how much would the tax be in problem 2?
   b. Can you figure out a way to calculate how much money will need to be put in the account now for the grandchild to have the $30,000 if the law does not pass and the tax has to be paid?

Section 4.2 Annuities: Future And Present Value

In the work of the last section you will note that we are depositing a fixed amount of money in some kind of an account and then letting it sit there to earn interest. More commonly, one would add some amount of money on some periodic basis to this account. When money is added to an account in a series of equal payments made at equal time periods the account is referred to as an **annuity**. The **future value** of an annuity is its total value, i.e. the sum of the payments and the interest earned on each payment.

Example 1: Annuities - future value

Suppose I agreed to pay you $8,000 per year at the end of the year for the next 5 years. What would be the value of this money at the end of 5 years? At first glance, you might think that $40,000 would be the correct answer. What would you do with the $8,000 that I gave you each year? Put it under the mattress? In all likelihood you would at the very least put it in some interest earning account. Thus, the actual amount of money you would end with would be more than the $40,000 that I gave you. Hence, in working this problem, we need to know how much money you would have at the end of the 5 years. This is known as "the future value of an annuity" problem.

In finding the solutions let us assume that you are able to get a yearly interest of r=6% on your money.

MATHEMATICAL FORMULA SOLUTION:

To derive the mathematical formula used to work this problem, we start by adding and simplifying the results and then we look for a pattern. Note that payments are still being made at **the end of each time period**. We will let EOY1, EOY2, EOY3, etc. represent the amount of money you would have at the end of each successive year.

End of year 1 = EOY1	End of year 2 = EOY2	End of year 3 = EOY3	etc.
$8000	8000+(EOY1+.06*EOY1)= 8000+EOY1(1+.06)^1	8000+(EOY2+.06*EOY2)= 8000+EOY2(1+.06)^1	etc.

End of year 1: $8000

End of year 2:   8000+EOY1(1+.06)^1

$$8000+8000(1+.06)^1$$

End of year 3: $8000+EOY2(1+.06)^1$
$8000+[8000+8000(1+.06)^1](1+.06)^1$
$8000+8000(1.06)^1+8000(1.06)^2$

End of year 4: $8000+EOY3(1+.06)^1=$
$8000+[8000+8000(1.06)^1+8000(1.06)^2](1+.06)^1$
$8000+8000(1.06)^1+8000(1.06)^2+8000(1.06)^3$

End of year 5: $8000+EOY4(1+.06)^1=$ DO YOU SEE A PATTERN?
$8000+8000(1.06)^1+8000(1.06)^2+8000(1.06)^3+8000(1.06)^4$

Since $8000 occurs in each term it can be factored out leaving:

$$8000(1+1.06+1.06^2+1.06^3+1.06^4)$$

This reduces our problem to one of finding the sum of $1+1.06+1.06^2+1.06^3+1.06^4$. To show you the general nature of what we are doing, let $1+r=1.06=x$. Then we wish to sum $S=1+x+x^2+x^3+x^4$. Such sums can be found with the following trick:

$$S = 1 + x + x^2 + x^3 + x^4$$
$$xS = x + x^2 + x^3 + x^4 + x^5$$

Hence, $S - xS = 1 - x^5$, (everything else cancels)

$$S = \frac{1-x^5}{1-x}$$ substituting $1+r$ back in for x, we get

$$S = \frac{1-x^5}{1-x} = \frac{1-(1+r)^5}{1-(1+r)} = \frac{1-(1+r)^5}{-r} = \frac{(1+r)^5-1}{r}$$

Since this formula sums the terms, we need only multiply by the payments, P, made at the end of each term to get the future value, F, of the annuity.

FUTURE VALUE of an ANNUITY at interest r with N PAYMENTS P per year for t years, each made at the <u>end</u> of the payment period:

$$F_a = P\frac{(1+(r/N))^{N*t}-1}{r/N}$$

The value of the annuity may now be found by evaluating the expression

$$8000*\frac{1.06^5-1}{.06} = 45096.74.$$

An important note of caution here:

Many students will memorize this formula but pay no attention to the derivation of it. If I had changed the problem only slightly to read that the $8,000 is paid at the start of each year at the end of the first year you would have 8000*(1.06) or 8000*x dollars instead of just 8000*1, ie.,

$$8000(1+1.06+1.06^2+1.06^3+1.06^4)$$ would be replaced by
$$8000(1.06+1.06^2+1.06^3+1.06^4+1.06^5).$$

This makes the first term of the expansion we sum an "x" instead of "1", $S = x + x^2 + x^3 + x^4 + x^5$, which causes our formula to be: $\frac{x - x^6}{1 - x} = x * \frac{(1 - x^5)}{1 - x}$. This formula is derived from using 5 payment periods with payments at the start of each period, so if we had n payment periods, upon substituting $1 + r = x$, the future value of the annuity at the **end of the last period** would be:

FUTURE VALUE of an ANNUITY at interest r with N PAYMENTS P per year, for t years each made at the <u>start</u> of the payment period:

$$_aF = P \frac{(1+r/N)^{N*t} - 1}{r/N} * (1+r/N)$$

Note that the only difference between the previous formula and this one is that the previous answer must be multiplied by 1+r=1+.06. Using the subscript "a" for annuity before F for payment made at the start and after F for payment made at the end.

Restatement of example 1:

Suppose I agreed to pay you $8,000 per year at the **start** of the year for the next 5 years with interest at 6%. What is this annuity worth at the end of 5 years?

$$_aF = 8,000 * \frac{(1+.06)^5 - 1}{.06} * (1+.06)$$
$$= (45096.74)(1.06)$$
$$= \$47,802.54$$

Again, we can store these formulas in the SOLVER of our calculator and use them to find **FUTURE VALUE** of an annuity. We can use "AF" (payment at the start of the period) and "FA" (payment at the end of the period) for our annuity formulas. Follow the steps indicated before to enter the following:

AF = P*(((1+(r/N))^(N*t) - 1)/(r/N))*(1+(r/N))
FA = P*(((1+(r/N))^(N*t) - 1)/(r/N))

Example 2:

Find the future value of an annuity with payments of $150 at a rate of 6.75% when the payments are made at the start of each month for 25 years. (Use AF)

$117,469.01

Example 3:

Find the future value of the same annuity in example 2 except the payments are made at the end of each month. (Use FA)

```
exp=P*(((1+(r/N))^(N...
 exp=116811.94484031
 P=150
 r=.0675
 N=12
 t=25
 bound=(-1E99,1E99)
GRAPH RANGE ZOOM TRACE SOLVE
```
$116,811.94

Example 4:

Find the future value of an annuity of $1800 made at the end of each year for 25 years. The rate is 6.75%. (Use FA)

```
exp=P*(((1+(r/N))^(N...
 exp=109843.76020174
 P=1800
 r=.0675
 N=1
 t=25
 bound=(-1E99,1E99)
GRAPH RANGE ZOOM TRACE SOLVE
```
$109,843.76

Example 5:

Find the future value of the annuity in example 4 except the payment is made at the start of each year. (Use AF)

$117,258.21

Example 6: Annuities, present value

Consider the problem of example 1 again, $8,000 for 5 years at 6% with payments at the <u>end</u> of each time period, except this time we wish to know the present value of this annuity. In other words, how much money would we have to deposit <u>right now</u> at 6% interest to end up with 45,096.74 in 5 years? From our previous work, we know that FA = $45,096.74.

1. You may work this problem by calculating the present value of $45,096.74 at a 6% annual (n=1) rate discounted back 5 years:

   $P = F\left(1 + \dfrac{r}{N}\right)^{-Nt}$

   P=45096.74*(1+.06)$^{-5}$=$33,698.91.

   Or we can use formula "F1" in the solver to solve for "P":

   ```
   exp=P*(1+(r/N))^(N*t)
    exp=45096.74
   ■P=33698.907534616■
    r=.06
    N=1
    t=5
    bound=(-1E99,1E99)
   GRAPH|RANGE|ZOOM|TRACE|SOLVE
   ```

2. We can derive a formula for finding the **PRESENT VALUE of an ANNUITY (PA) at interest r with N PAYMENTS of P per year for t years, each made at the <u>end</u> of the payment period.**

   From step 1 above replace "P" with "PA" (Present value of an annuity) and "F" with "FA" (Future value of an annuity) to get:
   PA = FA*(1+(r/N))$^{-N*t}$

   Now substituting for FA, $\mathbf{FA = P} * \dfrac{(1+(r/N))^{N*t} - 1}{(r/N)}$, this immediately yields

   an equivalent equation, $\mathbf{PA = P} * \dfrac{(1+(r/N))^{N*t} - 1}{(r/N)} * (1+(r/N))^{-N*t}$

   and simplification yields: $\mathbf{PA = P} * \dfrac{1 - (1+(r/N))^{-N*t}}{(r/N)}$

   Using our new formula:

```
exp=P*((1-(1+(r/N))^...
■exp=33698.910284526
 P=8000
 r=.06
 N=1
 t=5
 bound=(-1E99,1E99)
GRAPH|RANGE|ZOOM|TRACE|SOLVE
```
$33,698.91 as in method #1.

Unfortunately, just as before, this only works correctly when the payments are made at the end of each time period (yearly in this case). If payments are made at the **start of each time period**, working through all the algebra again yields only the following change:

$$AP = P * \frac{1 - (1 + (r/N))^{-N*t}}{r/N} * (1 + (r/N))$$

Using this new formula in the SOLVER:

```
exp=P*((1-(1+(r/N))^...
■exp=35720.844901597
 P=8000
 r=.06
 N=1
 t=5
 bound=(-1E99,1E99)
GRAPH|RANGE|ZOOM|TRACE|SOLVE
```
$35,720.84

In words, if you put $35,720.84 in the bank **right now** at 6% interest for 5 years, you would have **$47,802.54** at the end of the 5 years.

We can enter these formulas in the SOLVER as described previously:

PA=P*((1-(1+(r/N))^(-N*t))/(r/N)   PA- for payments made at the END of periods

AP=P*((1-(1+(r/N))^(-N*t))/(r/N)*(1+(r/N))   AP- for payments made at the beginning of periods

Terminology needed in Exercises 4.2

1. Future value of annuity equations:
   AF = P*(((1+(r/N))^(N*t) - 1)/(r/N))*(1+(r/N))   payments made at the beginning of the period

   FA = P*(((1+(r/N))^(N*t) - 1)/(r/N))   payments made at the end of the period

2. Present value of annuity equations:
   a. Method One
      Find the future value of the annuity using equations in number 1 above; then, find the present value (solve for P in the SOLVER) using equation F1 from section 4.1.

   b. Method Two
      PA=P*((1-(1+(r/N))^(-N*t))/(r/N)   payments made at the end of periods

      AP=P*((1-(1+(r/N))^(-N*t))/(r/N)*(1+(r/N))   payments made at the beginning of periods

3. Effective annual rate: What annual rate r equals the continuous rate k? Since $P(1+r)=Pe^k$ for one year, cancelling P we get $1+r=e^k$, or **$r=e^k-1$** **Note:** Do not enter this formula in the SOLVER as "r"; you might call it "er".

Exercise 4.2

1. Use your calculator to find

   a. the future value of an annuity with payments of $500 per month paid at the end of each month for 7 years with a guaranteed rate of 6% per annum.
   b. the present value of the annuity in a. (Show two methods)

2. Use your calculator to find

   a. the future value of an annuity with payments of $100 per month paid at the start of each month for 25 years with a guaranteed rate of 6% per annum.
   b. the present value of the annuity in a. (Show two methods)

3. Use your calculator to find

   a. the future value of an annuity with payments of $250 per quarter paid at the beginning of each quarter for 30 years with a guaranteed rate of 7.25% per annum.
   b. the present value of the annuity in a. (Show two methods)

4. Use your calculator to find

   a. the future value of an annuity with payments of $1000 semi-annually paid at the end of each period for 28 years with a guaranteed rate of 6.75% per annum.
   b. the present value of the annuity in a. (Show two methods)

5. If I agreed to give you $5,000 per year for the next 4 years to pay off a loan paying the $5,000 at the end of each year, how much money could I give you right now so that you would end up with the same amount if you put it in the bank at 7% interest compounded yearly?

6. If I paid you $2,000 per year at the start of each year for the next 5 years how much money would you have at the end of the five years if you put the money in an account that yielded 8% continuous interest? (Hint: Use effective annual rate.) How much money could I give you right now so that you would end up with this amount in 5 years if you could put it in an account that earned 7.25% interest compounded yearly?

7. If I paid you $3,500 per year at the end of each year for the next 4 years how much money would you have at the end of the four years if you put the money in an account that yielded 7.5% continuous interest? How much money could I give you right now so that you would end up with this amount in 4 years if you could put it in an account that earned 8% interest compounded yearly? See Hint above.

8. Two brothers Bill and Harry decide to save some money for retirement. Bill starts at age 21 and deposits $1000 a year through age 30, then does not deposit any more, leaving the amount on deposit to gather interest at 8% to age 65. Harry decides that having additional money for retirement is a good idea, so at age 31 deposits $1,000 a year to age 65 getting 8% on the money on deposit each year.

   a. How much money did each brother invest?
   b. Who ends up with the most money at age 65?
   c. Exactly how much money does each brother end up with?

9. Two brothers John and Paul decide to save some money for retirement. John starts at age 24 and deposits $2000 a year through age 40, then does not deposit any more, leaving the amount on deposit to gather interest at 8.25% to age 65. Paul decides that having additional money for retirement is a good idea, so at age 30 deposits $2,500 a year to age 65 getting 8% on the money on deposit each year.

   a. How much money did each brother invest?
   b. Who ends up with the most money at age 65?
   c. Exactly how much money does each brother end up with?

10. Suppose that the parents of a child set up an annuity paying 6.75% per annum for this child's college education. How much does each payment need to be if the payments are to be made monthly (end of month) for 18 years if they wish to have $45,000 for this education?

11. A rich relative dies leaving you $25,500. If this amount was the result of annual deposits (beginning of each year) of $4000 at 5% per annum, for how many years did he make these deposits?

12. A rich relative dies leaving you $50,000. If this amount was the result of monthly deposits (end of each month) of $250 at 7.25% per annum, for how many years did he make these deposits?

Section 4.3 Amortization, Calculating Payments

The equation $PA = P * \dfrac{1-(1+(r/N))^{-N*t}}{(r/N)}$ can be used to calculate the amount of each payment to retire a loan in N*t payment periods. We need only interpret PA as the amount of the loan L you get today and P as the size of each payment you will need to make at the end of each payment period. Solving for P we get:

$$P = L * \dfrac{(r/N)}{1-(1+(r/N))^{-N*t}}$$

Amortization is the process of repaying a loan using payments that are part principal and part interest when paid by a sequence of equal periodic payments.

Example 1:

For example, suppose you borrowed $8,000 at 12% and needed to repay it in equal monthly payments over a 4 year period. Now, L=$8,000, .12/12=interest rate/12, and N*t=48 (number of payment periods) = (4 years)*12 (months in a year), and P=payments=? Substituting these values in our equation we get:

$$8000 * \dfrac{\dfrac{.12}{12}}{1-(1+\dfrac{.12}{12})^{-48}} = \$210.67.$$

We can store this in the SOLVER as:

$LP = L * \dfrac{(r/N)}{1-(1+(r/N))^{-N*t}}$ LP for loan payment

Example 2: (TI-82 users see Calculator Instructions III Section 4.3 for TI-82)

Find the monthly payment and the total interest paid, for a loan of $5000 at a rate of 11.75% per annum, if the loan is to be paid back in

a. 3 years
b. 4 years

Solution:

a.  monthly payment of $165.48

Total interest paid = [$165.48 * (12*3)] − $5000 = $5957.28 − $5000 = $957.28

b.  monthly payment of $131.06

Total interest paid = [$131.06 * (12*4)] − $5000 = $6290 − $5000 = $1290

The graph of the monthly payment (y1) vs. t in years (x-VAR) and the total interest paid, letting LP be y1 and t be x-VAR in the GRAPH editor, is as follows:

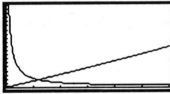

Using the TABLE program, we can look at the following:

x	y1	y2
1.00	443.66	323.91
2.00	234.78	634.82
3.00	165.48	957.11
4.00	131.06	1290.70
5.00	110.59	1635.50
6.00	97.10	1991.35

We can see from this that if the loan is to be paid back in monthly payments for 6 years, the monthly payment will be $97.10 and the total interest paid over the 6 years is $1991.35.

Example 3:

If the monthly payment of a $50,000 mortgage is $384.46 with 8.5% annual interest, find the term of the loan "t".

Using LP in our SOLVER,

Note the bound. The term is 30 years.

Example 4: Amortization Schedule

A car cost $15,000. After a down payment of $3,500, the balanced will be paid off in 48 monthly payments with interest of 5.75% per year on the unpaid balance.

a. Find the total amount of each payment.
b. Find the total amount of interest paid.
c. Find the part of the first payment that is interest and the part that is applied to reducing the debt.
d. Use the TABLE Program to generate a table showing the balance owed on the loan after each monthly payment.

Solution:

a. $15,000 − $3500 = $11,500 is the amount of the loan.
   Use LP in the SOLVER to get a monthly payment of $268.76

b. $268.76 * 48 = $12900.48
   $12900.48 - 11,500 = $1400.48

c. The monthly interest rate is $i = .0575/12$
   During the first month the entire $11,500 is owed. The interest (I) on this amount for 1 month is I = (Principal)(i)(payment #) = (11,500)(.0575/12)(1) = $55.10

   $55.10 is the interest in the first payment
   $268.76 − $55.10 = $213.66 is the amount of reduction of the loan.

d. Set $y2 = 268.76167857$ (monthly payment) in the GRAPH editor
       $y3 = .0575/12$ (monthly rate)
       $y1 = y2((1 − (1 + y3)^{-(48-x)})/y3$ (x = the payment #)
              we treated this as present value of an annuity

x	y1	x	y1	x	y1
0	11499.9999996	6	10202.5998294	12	8867.44971336
1	11286.3424877	7	9982.72560837	13	8641.177898
2	11071.6612002	8	9761.79782334	14	8413.82186352
3	10855.9512316	9	9539.811426	15	8185.37641471
4	10639.2076527	10	9316.76134385	16	7955.83633146
5	10421.4255108	11	9092.64248005	17	7725.19636865

x	y1	x	y1	x	y1
18	7493.45125601	24	6079.47410235	30	4624.3550079
19	7260.59569804	25	5839.84323719	31	4377.75169707
20	7026.62437385	26	5599.0641408	32	4129.96674538
21	6791.53193707	27	5357.13131123	33	3880.9944908
22	6555.3130157	28	5114.03922019	34	3630.82924416
23	6317.96221199	29	4869.78231289	35	3379.46528905

x	y1	x	y1	x	y1
36	3126.89688166	42	1585.86780137	48	0
37	2873.11825064	43	1324.70507268	49	-268.761678572
38	2618.12359702	44	1062.29093924	50	-538.811173518
39	2361.90709402	45	798.619404756	51	-810.15465563
40	2104.46288694	46	533.684444165	52	-1082.79832526
41	1845.78509303	47	267.480003555	53	-1356.74841248

Terminology needed in 4.3

1. Loan payment equation

$$LP = L * \frac{(r/N)}{1 - (1 + (r/N))^{-N*t}}$$

Exercise 4.3

1. Determine the monthly payment and the total interest paid for a loan of $12,000, at a rate of 10.5% for the given t in years: (Use the UT table program)

   a. t = 6   d. t = 9   g. t = 12
   b. t = 7   e. t = 10
   c. t = 8   f. t = 11

2. Calculate the difference in total payments (made monthly) between taking out a $180,000 loan for 20 years vs. 30 years at 8.5% annual interest.

3. Calculate the difference in total payments (made monthly) between taking out a $125,000 loan for 25 years vs. 30 years at 9.25% annual interest.

© 1995 Saunders College Publishing

4. You decide to buy a home for $100,000. The bank agrees to grant your loan for 30 years at 8% interest. You will repay the loan in monthly payments made at the end of each month.

   a. Find your monthly payments.
   b. How much money do you end up paying for the home?
   c. What is the total interest for the term of the loan?

5. If the monthly payment on a $110,000 mortgage is $845.80, with a rate of 8.5% per annum, find the term of the loan.

6. A new car at Dealer A is priced at $12,000; you must pay $3500 down and pay the balance in 48 monthly payments at an annual interest rate of 8.5% compounded monthly. At another Dealer B, the car costs $11,700, with the same down payment and 48 monthly payments at an annual interest rate of 11.75% compounded monthly. Which is the better deal?

7. When John Doe opened his new business, he bought $18,000 worth of inventory and $5,000 worth of shop tools. He paid $3,000 down and agreed to pay the balance in semi-annual payments for 6 years at 12% interest. Prepare an amortization schedule showing the balance at the end of each pay period.

8. The Clark family bought a home for $96,500. They paid $14,000 down and financed the balance at 9.5% for 30 years. Find the balance on the loan after the 120th payment.

Section 4.4 Additional Differentiation (ln x and e^x)

From the standpoint of calculus our interest is in the occurrence of this new function, e^t. We see that it crops up in a very natural fashion when we try to calculate continuous interest. How does one differentiate such a function? Returning to our definition of slope function, we see that we need to calculate

$$\lim_{h \to 0} \frac{e^{x+h} - e^x}{h}.$$ This fraction can be factored as follows:

$$\lim_{h \to 0} \frac{e^{x+h} - e^x}{h} = \lim_{h \to 0} \frac{e^x e^h - e^x}{h} = \lim_{h \to 0} e^x \left(\frac{e^h - 1}{h} \right).$$

Thus we can see that evaluating the derivative of this function comes down to evaluating the limit: $\lim_{h \to 0} \frac{e^h - 1}{h}$. This is not an easy thing to do, so we will refer to our mathematician friends who tell us that this limit is 1.

This yields a most surprising result, namely:

Elementary derivative number 2: If $y = e^x$, then $y' = e^x$.

The function e^x has the interesting property that it is its own derivative! For example, suppose you wanted to know the slope of e^x when x=2. Both the value of the function and the value of its slope is e^2.

Recalling that the theorems we have developed work for arbitrary functions, we may apply the product theorem, chain rule theorem, etc. to get the following results immediately:

Example 1: If $f(x) = xe^x$, then $f'(x) =$ (product rule) $e^x + xe^x$
Example 2: If $g(x) = e^2$, then $g'(x) = 0$. (The derivative of a constant is 0)
Example 3: If $h(x) = 2e^x$, then $h'(x) = 2e^x$.
Example 4: If $f(x) = e^{7x}$, then $f'(x) =$ (chain rule) $e^{7x}(7) = 7e^{7x}$.

This shows why everyone(?) likes to differentiate e^u.
To differentiate e^u you copy the entire function down and multiply by the derivative of the exponent.

Example 5: If $h(t) = e^{-.08t}$, then $h'(t) = e^{-.08t}(-.08) = -.08e^{-.08t}$.

Example 6: If $p(q) = q^2 e^{.1q}$, then $p'(q) = 2qe^{.1q} + q^2 e^{.1q}(.1)$.

and finally, because of the nature of exponents you can do some things with exponential equations you cannot do with other equations. For example, it is sometimes possible to avoid the use of the chain rule.

Example 7: If $f(x) = (e^{5x})^2$, you could calculate the derivative by using two applications of the chain rule, but if you rewrite this as $f(x) = e^{10x}$, one application will do. Recall the difference between $(e^x)^2$, which can be rewritten as e^{2x} and has derivative $2e^{2x}$, and e^{x^2} which cannot be rewritten and has derivative $2x\, e^{x^2}$.

Recalling from your algebra days, given any number we can define a function "logarithm" to that base. Hence, given this number e, we can define a function, $\log_e x$, for any number x>0. Since some of you may have forgotten your work in logarithms (or may not have understood it at the time), let's take a minute out for a quick review of logarithms.

Consider the following table.

number	powers of 10 needed	powers of e needed
-100	none	
-10	none	
0	none	
1/100	10^{-2}	
1/10	10^{-1}	
1	10^0	
10	10^1	
100	10^2	

First, remember that the word logarithm means exponent, a particular exponent. Hence, in this first table, the **logarithms base 10** for the given numbers are simply the powers of ten you need to get those numbers.

Hence, the $\log_{10} 100 = 2$, whereas the $\log_{10} \dfrac{1}{100} = -2$. Note that you cannot find the logarithm for any number less than or equal to zero, since negative exponents yield fractions, and you can get a fraction as close to 0 as you want, but you cannot get 0.

What I would like to do with you now is to guess some values we might need as exponents of e to get these same numbers. Now remember, we are going to guess since this will help to improve your concept of what the value of e is. In fact, there is one exponent we do not need to guess. You know that $e^0 = 1$, so we know one logarithm right off: $\log_e 1 = 0$.

Recall that $e = 2.71828\cdots$, so it has a value of about 3. We now guess the exponent that gives 10. Clearly it will be more than 2 since 3^2 is 9, so e^2 will be even less than 9. We can see that it will also have to be less than 3, for 3^3 is 27 considerably larger than 10, so e^3 should also be larger than 10. I will guess 2.5 and to check it, enter 10 on your calculator and press the $\log_e$ key. On most calculators, the $\log_e$ key is designated by ln or ln(x). On some calculators, there may be only one log key, and in this case it is most likely $\log_e$.

Checking our guess, we see that the correct answer is about 2.3. Now note that if the correct answer for $\log_e 10 = 2.3$, then the correct answer for $\log_e = -2.3$. To find the proper exponent of e to get 100, we again observe that since e is about 3, and 3^4 is 81, 3^5 is 243, we would guess the power of e to be between 4 and 5. A quick check on your calculator or computer yields $e^{4.6} = 100$. Hence, $\log_e 100 = 4.6$ and similarly, $\log_e = -4.6$. Looking in the table we see the correct values listed.

number	powers of 10 needed	powers of e needed
-100	none	none
-10	none	none
0	none	none
1/100	10^{-2}	$e^{-4.6}$
1/10	10^{-1}	$e^{-2.3}$
1	10^0	e^0
10	10^1	$e^{2.3}$
100	10^2	$e^{4.6}$

A little note your last instructor may not have told you. When reading $\log_b x = y$, it will be helpful if you will **not** read that as it stands. Read it as:

$\log_b x$ = What power of b do you need to get x? The answer is y.

In so doing you will resolve some seemingly hard questions in a very simple fashion. For example, $\log_3 x = 2$, solve for x. The solution is found by simply reading it: "what power of 3 do you need to get x? The answer is 2," ie, 3^2 is the value of x. This statement goes along nicely with the basic definition of $\log_e$, namely:

$$y = e^x \text{ if and only if } \ln(y) = x$$

Continuing with our calculus concepts we now need to know the derivative of ln(x). A clever trick can be used here to figure out the derivative. It is not important to business majors so we will not use it, but it involves the above definition and implicit differentiation. If you are really curious, ask your instructor to do it after class since not only does every math instructor know how to do it, they will be **delighted** to show you. The result will be:

Elementary derivative number 3: The derivative of ln(x) is $\dfrac{1}{x}$.

Example 8: Find the derivative of $h(x) = \ln(x^2+1)$.

solution: This is another application of the chain rule. You may view the function as:

1. h(u) = ln(u), and
2. u(x) = x²+1

Since h'(u) = and u'(x) = 2x, $\dfrac{d}{dx} \ln(x^2 + 1) = \dfrac{1}{x^2 + 1}(2x)$.

Example 9: Examining advertising expenditures

When spending money on advertising one needs to be aware of the limited affect the expenditure has on the sales. Usually, sales will have an increase while the advertising campaign is in effect but only up to a point. At some point, additional expenditures on advertising will not be justified by any significant return in sales. Is there a function which has this property, ie., it increases rapidly up to a point, and then fails to increase much further? There are several such functions. The property they would all have to have in common is a horizontal asymptote at some advertising cost.

Consider the function $y = 1 - e^{-kx}$, where x=number of dollars spent on advertising and y=sales response to the dollars spent.

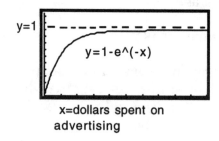
x=dollars spent on advertising

You can see that this function at least appears to have the necessary characteristics. That is:

1. If x=0 (no money is spent on advertising), the result is y=no sales, and
2. If you increase the advertising budget beyond some point little gain in sales are realized. What is the optimum strategy?

To resolve this question we must put in some actual numbers. Suppose records indicate that the maximum sales you can expect are 200 units per week. Suppose also you have found from previous experience that the correct function is

$$q = 200(1 - e^{-.01x})$$

Your profit on each item is $50. What is the best amount to spend on advertising?

solution: Let q be the quantity sold per week. Then $q = 200(1-e^{-.01x})$, where x is the amount you spend on advertising. Since your profit must have the amount you spent on advertising subtracted from it, Profit = P = 50q-x which yields $P = 10,000(1-e^{-.01x}) - x$ when we substitute for q.

Looking at a graph of the function we see that this is a classical calculus-type problem.

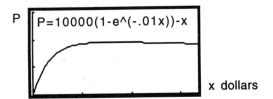
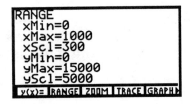

There is a maximum and it occurs at a peak of the function. Hence, we must differentiate, set the derivative equal to 0 and solve for x.

Differentiating $\frac{d}{dx}[10,000(1-e^{-.01x})-x]$ we get:

$\frac{dP}{dx} = 100e^{-.01x} - 1$ and solving $100e^{-.01x} - 1 = 0$ we get $e^{-.01x} = 1/100$. Taking the natural log of both sides, $\ln(e^{-.01x}) = \ln(1/100)$ and applying the definition of logs: $\log_e e^{-.01x} = -.01x$ and the division property: $\ln(1/100) = \ln(1) - \ln(100) = -\ln(100)$ this yields $-.01x = -\ln(100)$. Hence,

$$x = \frac{\ln(100)}{.01} = 100 * \ln(100) = 460.517.$$

Thus, at an advertising level of x = $460.517, maximum profit of $9,439.50 is realized.

One additional comment is in order here. The function we chose, $y=1-e^{-kx}$, was chosen for simplicity. Note that when x=0 which represents no advertising expenditures, sales are 0. This is not usually the case. A more general form of this function would allow for initial sales volume v with no advertising. This form would be:

$$y=c(a-e^{-kx})$$

where c(a–1)=v would be the initial sales volume before advertising (x=0). After you grasp these ideas a little better, we will use this more general form to solve a few problems.

Terminology needed in exercise 4.4

In this exercise set you will need to use all of the theorems you have had to date including the product rule, the quotient rule and the chain rule.

1. If $y = e^t$, then $y' = e^t$.
2. The derivative of $\ln(x)$ is $\dfrac{1}{x}$.
3. Basic properties of logarithms:
   a. $\ln(xy) = \ln(x) + \ln(y)$
   b. $\ln(x/y) = \ln(x) - \ln(y)$
   c. $\ln(x^n) = n*\ln(x)$
   d. $\ln(x) = y$ means: What power do you have to raise e to, to get x? The answer is y, ie., $\ln(x) = y$ means the same as $e^y = x$.

Exercise 4.4

In 1-10 find the derivatives.

1. e^{3x}
2. $\ln 2x$
3. e^{3x+1}
4. $\ln(2x^3-1)$
5. $3xe^{2x}$
6. $2x\ln(2x)$
7. $e^{4x}\ln(4x)$

8. $\dfrac{e^t}{2t}$

9. $\dfrac{5xe^{x^2}}{x^3+1}$

10. $3\dfrac{\ln(2x)}{5x}$

Now let's consider some "real life" problems. It is usually the case that when students see any of the following functions their response is: "you must be kidding!" Unfortunately for you, the kinds of functions you have been looking at in your math books until now are the ones about which you have been kidded. Such functions as quadratic equations either do not occur in business, or, if they do, a good businessman with a little data can guess the answer a lot easier than using math to find it. However, the following types of functions do occur and are difficult or impossible to deal with without the training and tools which you have just been given. So.....welcome to the real world.

11. When a certain textbook is first published, its sales rate (derivative of the sales function) roughly follows a curve (in 1000's per year) given by the equation

$$\text{Rate}(t): \quad R(t) = 100 \dfrac{t^2 e^{\frac{-t^2}{2}}}{3!}, \quad t>0, \ t \text{ in years}$$

To interpret this curve correctly, each value of Rate(t) on the curve should be interpreted as the rate of sales at the time calculated. For example, at the end of the first month, t=1/12 and R(1/12)=0.115339 = the rate of sales at the end of the first month. If this rate of sales (0.115339) were to continue for the next year, we would sell 0.115339 thousands of books or about 115, a sad state of affairs. Fortunately, R(1)=10.1088 which means that by the end of the first year sales have increased to a yearly rate of 10,109 books per year.
NOTE: "!" is found on the calculator in MATH (2nd X), PROB, F1

   a. Graph the curve for 0≤t≤4.5
   b. Calculate the derivative of the sales rate.
   c. Graph the derivative.
   d. Make some observations about the sales rate with respect to where the values of the derivative are
      1. positive
      2. negative, and
      3. zero.
   e. When does the maximum sales rate occur, what is it, and label this point on your graph.

f. When is the sales rate below 5,000 per year (two values)?

12. When a textbook is first published, if it is to be successful the sales rate roughly follows a curve (in 1000's per year) given by the equation

$$\text{Rate}(t): R(t) = v \frac{t^2 e^{\frac{-t^2}{2}}}{3!}, \quad t>0, \text{ t in years, v to be calculated.}$$

To interpret this curve correctly, each value of Rate(t) on the curve should be interpreted as the rate of sales at the time calculated. For example, at the end of the first month, t=1/12 and R(1/12)=the rate of sales at the end of the first month. If this rate were to remain constant, the number you get for R(1/12) would represent how many thousands of books you would sell during the next 12 month period. Similarly, R(1) would be the sales rate at the end of the first year. Suppose it is observed that the rate of sales Rate(t)=100 at the end of the first month. Use the SOLVER to find the value of v. Then:

a. Graph the curve for $0 \leq t \leq 4.5$
b. Calculate the derivative of the sales rate.
c. Graph the derivative.
d. Make some observations about the sales rate with respect to where the values of the derivative are
   1. positive
   2. negative, and
   3. zero.
e. When does the maximum sales rate occur, what is it, and label this point on your graph.
f. When does the sales rate fall below 5,000 per year (two answers)?

13. One important class of functions are those of the form $y = ae^{-b(x-c)^2}$. These functions represent the normal or bell-shaped curve. The normal curve represents such diverse things as the number of people with size 7 shoes and the number of defective parts turned out by a machine tool.

a. Calculate the x value where the maximum of all such curves occur.
b. From your last conclusion, without calculating, state the x value where the maximum of
$$y = 7e^{-(x-3)^2}$$ will occur.

Now graph and label the maximum on the following curves:

a. $7e^{-(x-3)^2}$

b. $.5e^{-(x-3)^2}$

c. $e^{-(x-10)^2}$

d. $e^{-5(x-3)^2}$

14. Suppose your sales of a collection of items yielding an average profit of $4 (includes other cost but NOT advertising cost) are about 600 per week. You think that with some advertising the sales will increase. You buy a varying number of advertising spots and observe the change in sales. The relationship between advertising costs, x, and sales quantity, q, is roughly $q=300(2-e^{-.02x})$. You wish to find out about what the optimal amount of money you should spend on advertising should be.

   a. Graph the profit function $P = 4(300(2-e^{-.02x})) - x$.
   b. Calculate optimum expenditures on advertising, find the profit at this value and label this point on your graph.
   c. At this value, is profit very sensitive to advertising expenditures, ie., how much difference does an increase of $100 make in profit? How about a decrease of $100?

Section 4.5 When The Derivative Fails: Using Limits To Find Peaks And Valleys

It is frequently the case that looking only at the derivative of a function to find where the function has a maximum or a minimum can be misleading and even incorrect. Most of these problems can be eliminated using a clever combination of calculus, graphing, and limits. In all previous problems, our relative maximum or minimum has occurred at the same place that the derivative had a root. Be forewarned that there are plenty of real world problems where there is a maximum or minimum, but finding the roots of the derivative will provide no help in locating them. A few examples will illustrate this point.

Example 1: Gompertz Model: Using calculus, graphs, and limits

Many exponential models have maximums that, while easily found by using limits, are not found by using calculus. One of these is the Gompertz model, $\mathbf{y = k * a^{b^x}}$.

For example $y = 683 * .053^{.8667^x}$ is used to represent the growth of ice cream sales for 1929-1996. Differentiation of this function yields no useful information.

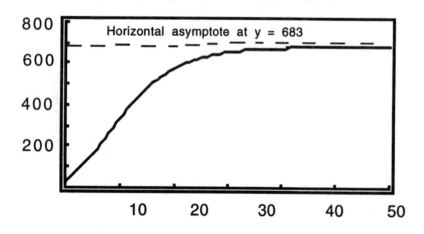

Note that the function approaches but never quite reaches the value 683. When this happens, the value 683 is referred to as a **horizontal asymptote**. Since this is an exponential function, why don't the values of the function keep getting larger as x increases? The answer is found when you examine the exponent .8667x and see that .8667 is a decimal smaller than 1. Raising this to higher powers yields smaller values. Consider the fraction $\left(\frac{1}{2}\right)^x$. As x increases you get smaller and smaller values for the expression since $\left(\frac{1}{2}\right)^2 = \frac{1}{4}$, $\left(\frac{1}{2}\right)^3 = \frac{1}{8}$, etc. Thus,

$$\lim_{x \to \infty}(683 * .053^{.8667^x}) = 683 \lim_{x \to \infty}(.053^{.8667^x}) = 683 * (.053^0) = 683$$

Example 2: The modified exponential growth curve

Another example of a situation where differentiation is of little use is on a modification of an exponential growth curve, $y = k(a - be^{-cx})$.
For example $y = 1000(1 - e^{-x})$ is used to predict the response of sales to an advertising campaign.

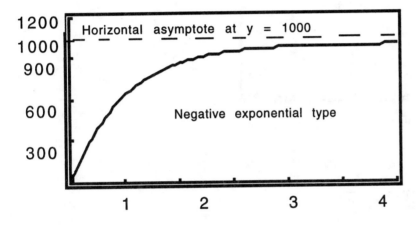

Using our limit approach and recognizing that $e^{-x} = \dfrac{1}{e^x}$, we arrive at the following maximum value for the function:

$$\lim_{x \to \infty} 1000(1 - e^{-x}) = 1000 * \lim_{x \to \infty}(1 - e^{-x}) = 1000 * \left(1 - \lim_{x \to \infty} e^{-x}\right) = 1000 * \left(1 - \lim_{x \to \infty} \frac{1}{e^x}\right)$$

$$= 1000 * (1 - 0) = 1000$$

thus we see that there is in fact a maximum and it occurs at 1000.

Example 3: Logistic models

Next consider the logistic curve, $y = \dfrac{k}{1+e^{a+bx}}$. This curve is frequently used to predict population growth. As a specific example consider $y = \dfrac{1000}{1+e^{1.18-.13x}}$. Calculating the derivative is of little use. However, you can see that the curve flattens out at 1000.

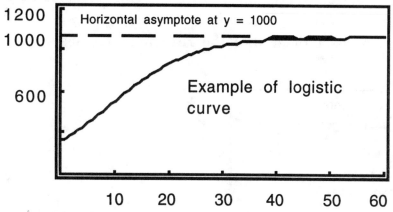

Example of logistic curve

Taking the limit

$$\lim_{x\to\infty}\left(\dfrac{1000}{1+e^{1.18-.13x}}\right) = 1000 * \lim_{x\to\infty}\dfrac{1}{1+e^{1.18-.13x}}$$

and examining the $\lim_{x\to\infty} e^{1.18-.13x}$ we get $\lim_{x\to\infty}\dfrac{e^{1.18}}{e^{.13x}}$ and as $x \to \infty$ $e^{.13x}$ becomes very large, making $\dfrac{e^{1.18}}{e^{.13x}}$ very small since $e^{1.18}$ is a constant. Therefore,

$$\lim_{x\to\infty}\left(\dfrac{1000}{1+e^{1.18-.13x}}\right) = 1000 * \lim_{x\to\infty}\dfrac{1}{1+e^{1.18-.13x}}$$

$$=1000\left(\dfrac{1}{1+0}\right)=1000.$$

Example 4: Normal curves

Finally, consider a normal curve, $y = ae^{-b(x-c)^2}$, one of the most commonly used curves. As a specific example, consider $y = e^{-(x-3)^2}$. Differentiating yields $y = -2(x-3) * e^{-(x-3)^2}$. Finding the roots of the derivative yields x=3.

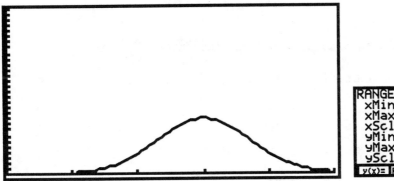

As we can see from the graph, this gives the location of the maximum. Where is the minimum and what is the value? Again from the graph it appears that the minimum is y=0 in both directions. This can be easily calculated by using limits:

$$\lim_{x \to \pm\infty} e^{-(x-3)^2} = \lim_{x \to \pm\infty} \frac{1}{e^{(x-3)^2}} = 0$$

Example 5: Pollutant removal

Consider the removal of pollutants. The cost of removal generally increases as the percentage you wish removed is increased. Suppose for a certain type of pollutant this relationship is given by $p(c) = \dfrac{100c}{c + 12000}$ $c \geq 0$.

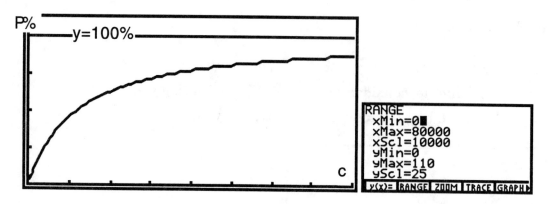

This function clearly has a horizontal asymptote at p=100%. To calculate this algebraically one needs to calculate $\lim_{c \to \infty} \frac{100c}{c+12000}$.

Taking the limit as it is yields $\frac{\infty}{\infty}$. This difficulty can easily be overcome by dividing the top and bottom of the expression by c, yielding $\lim_{c \to \infty} \frac{100}{1+\frac{12000}{c}}$, and this value is 100%. Hence, both algebraically and graphically we see that there is a horizontal asymptote at p=100.

In addition, the **derivative** of this function has a horizontal asymptote. To understand this, consider what happens as you substitute ever larger cost values into the pollution percentage equation. As you do, you will discover that the relative return decreases. For example, at the level of spending of $5,000, an increase in spending of $5,000 yields significant results: p(5000) = 29% and p(10000) = 45% pollutant removal, whereas at the $50,000 level an increase of $5,000 yields insignificant results: p(50000) = 81% and p(55000) = 82% removal.

Clearly, the effectiveness of the money you spend is decreasing. In fact, using the derivative of p(c), $p'(c) = \frac{1200000}{(c+12000)^2}$, you can easily see that as cost c increases, the value of the denominator increases rapidly thus rapidly decreasing the value of the fraction. Hence, the derivative shows that the **rate of return** on your investment decreases rapidly above $12,000. This is not a happy state of affairs but does reflect the world we live in.

Example 6: Asymptotes for discrete data

Now consider the case where you have a collection of data points.

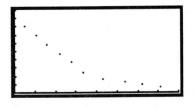

Discrete Data Points - Horizontal Asymptote is the x-Axis

If you were to proceed in an ignorant fashion, you might fit a curve (like a quadratic) to this collection of data points. If you were to do this, you would be setting yourself up for many possibly erroneus conclusions.

First, your choice of function must be determined by the conditions governing the data itself. In this example, I have not given you any background information. This makes fitting any curve a potentially dangerous operation.

Second, if you do fit the wrong curve and then apply mathematics, all of your results will be invalid. While you may not realize it at the moment, this is done more often than you think. The real danger lies in the fact that you can generate enough numbers and do enough mathematics with what you have to make your work believable.

IMPORTANT SUMMARY:

It is frequently the case in business problems that the maximum of a function occurs **at a horizontal asymptote,** not at a peak. Generally, the student can find this out **only by looking at a correctly scaled graph**. When this does occur, one **does not find the maximum by calculating the root of the derivative. Limits must be used**.

At this point, you need to work the following problems by yourself. They are relatively easy once you get a graph and decide which of the two procedures,

1. roots of derivatives, or
2. limit of the function

must be used.

Information needed to work exercise 4.5

1. Understand the meaning of "limit" and in what circumstances they are to be used.
2. Ability to do simple limits.

Exercise 4.5

In problems 1-11 graph each function for $x > 0$, then find all maximums and minimums. Use both differentiation and limits as needed.

1. $G(x) = 512 * .04^{.9^x}$, Gompertz curve

2. $f(x) = 750(2 - e^{-x})$, modified exponential growth curve

3. $L(x) = \dfrac{1000000}{1 + e^{1.18 - .13x}}$, logistic curve

4. $h(x) = \dfrac{3x}{1 + x}$

5. $h(x) = 683 * .053^{.8667^x}$, (Gompertz curve fitting ice cream production, 1929-1996)

6. $f(x) = 1000(1 - e^{-x})$, (modified exponential growth curve predicting sales response to advertising)

7. $h(x) = \dfrac{1000}{1 + e^{1.18 - .13x}}$, (logistic curve predicting population growth)

8. The fixed cost of producing a new television is calculated to be $100,000 and the variable cost is $390.
   a. Graph the average cost function and calculate both graphically and algebraically what value the average cost ultimately approaches.
   b. What is the break-even point if the company sells the sets at $425 each?

CHAPTER 5: APPLYING CALCULUS CONCEPTS TO DISCRETE DATA

Section 5.1 Definition Of Continuity

You may think of a continuous function as a function which is drawn without lifting the pencil from the paper. It is true that in most business situations one is forced to deal with a collection of discrete points, or at best a function which has steps and gaps in it. Such functions are clearly not continuous. However, it is often the case that the function is continuous in pieces. By this, I mean that it is continuous for a piece, then there may be a jump or gap, then continuous for another piece.

Even when you only have a collection of discrete points you can think of this set as possibly having an underlying function which is continuous within each interval over a collection of intervals, or which is a continuous function passing through the discrete points we have even though we cannot find exactly what that function is.

It may be that the function is totally or partially unpredictable (ie. the stock market), in which case, there may or may not be an underlying continuous function but in any case we will never find it.

This is the bad side of considering continuous functions in business situations. On the bright side, we can often come up with a continuous function (or a collection of continuous functions) which exactly (or nearly) goes through all the points we have. Using our imagination we can often apply advanced techniques learned when dealing with continuous functions to analyze discrete sets of data. For example, you already know the value of looking at the points where the slope of a (continuous) function is zero. This can be useful, and, in fact, you can always fit a continuous curve through a discrete set of data points.

For example, in the figures below, on the left we have a set of data points. On the right we have a continuous function that goes through those points.

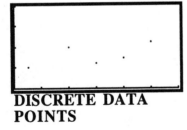
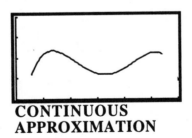

DISCRETE DATA POINTS   **CONTINUOUS APPROXIMATION**

The problem with doing this is that there are many functions that will go through all the given data points and the selection of the correct one can be difficult or impossible.

As can be seen in this diagram, the correct function may not have nice smooth curves connecting the points.

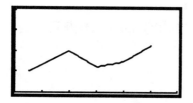 **PIECEWISE LINEAR**

This situation is called "piecewise linear" since the line connecting each successive point is a straight line segment. Piecewise linear functions are the type we use in an xyLINE graph in the STAT editor of our calculators. In terms of calculus, these functions cause something of a problem since they are continuous but their derivatives are not.

If you look at this function and think of what the derivative will look like, you can see the slope of the line joining each successive pair of points changes abruptly. Hence, a graph of the derivative would look like this:

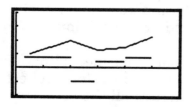 **PIECEWISE LINEAR FUNCTION AND DERIVATIVE**

In addition, it is often the case that the function we should consider through the given data points is only continuous in pieces. A simple example of this is the accompanying function, often called a "step function" for the obvious reasons. Notice that the derivative of this function does exist everywhere because the function is always a constant and the derivative is zero.

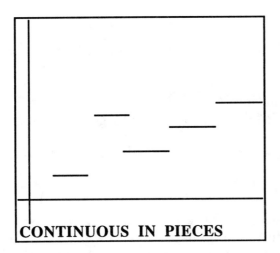

A simple example of a step function would be the discounted prices charged for quantity buying, eg., $20 each for 1-100 items, $18 each for 101 to 500 etc. Such functions are not only common in business, but prevalent. It is a rare situation indeed where volume purchasing does not yield an additional discount.

I recall the time a company I was working with was using plastic to laminate sheets of paper on both sides. We decided to see if volume purchasing would significantly reduce our cost. At the time we were paying about 13¢ a square foot wholesale. We contacted the manufacturer and by ordering their minimum quantity we cut the price to 2¢ a square foot.

Just so you won't think it was all that wonderful, their minimum order size was a railroad car full of rolls of laminate. We had to get a railroad car AND a siding to park it on. After we did this, we did realize tremendous saving and were able to cut our price to the consumer from 23¢ a square foot of finished work to 14¢.

Even though we will not ordinarily get a nice continuous function to work with, just having continuous pieces can be an advantage when doing calculus. We have for some time been dealing with both continuous and discontinuous functions without specifically saying what they are. We really need a good definition to avoid drawing erroneous conclusions about continuous functions (even if they just occur in pieces).

Rather than the "don't lift the pencil off the paper" definition, let's consider the following accurate definition of a continuous function:

CONTINUOUS FUNCTION:

A function "f" defined on the real numbers about some point "a" is continuous at a point "a" if

1. f(a) exists (you can evaluate function "f" at "a"),
2. $\lim_{x \to a} f(x)$ exists (as you approach "a" from either side you get the same value), and
3. $\lim_{x \to a} f(x) = f(a)$ (the value you get when you get there is what you would have gotten by just calculating f(a)).

If you have a function f(x) and wish to check for continuity, your first effort should be directed towards graphing the function. Looking at a graph will not assure you the function is continuous, but seeing a huge gap between two pieces will certainly help convince you it is not! In addition, some functions such as those with volume discounts, may appear discontinuous at first, but the graph will look smooth enough to make you look further.

Example 1: Linear equations

If you have any linear equation, f(x)=mx+b, the appearance of the graph will tell you that the function is continuous. You may also apply the 3 conditions given above as follows:

1. f(a) exists (you can evaluate function "f" at "a") since f(a) = ma+b.

2. $\lim_{x \to a} f(x)$ exists (as you approach "a" from either side you get the same value) since nothing peculiar occurs, and

3. $\lim_{x \to a} f(x) = f(a) = ma+b$ (the value you get when you get there is what you would have gotten by just calculating f(a)).

Example 2: Polynomials

Polynomials are continuous functions. To see this, you need only consider the 3 conditions applied to the term x^n. Again, nothing of an unusual nature occurs when you approach any value for x.

Example 3: Rational Functions
Quotients-vertical asymptote

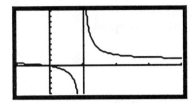 $f(x) = \dfrac{x}{x-1}$, vertical asymptote at x = 1

Only a moment is required when graphing functions such as $f(x) = \dfrac{x}{x-1}$ to convince you that fractions in general will not satisfy the 3 conditions. In this example, condition 1 is not satisfied at the point x=1 since the function is not defined at that point. It should be clear to you that any quotient which has a denominator that takes on the value 0 can be a problem. While this is not always the case (mathematicians will find exceptions to almost anything!) it is the case most of the time and graphing the function will reveal most difficulties.

Quotients-point discontinuity

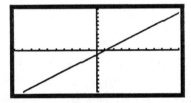 $f(x) = \dfrac{x^2 - 1}{x+1}$, point discontinuity at x=-1

NOTE: Point discontinuities with quotients occur where that value of x causes the denominator to take on the value of zero and **is** a factor of the numerator. Vertical asymptotes occur where the value of x causes the denominator to take on the value of zero and it **is not** a factor of the numerator. Graphics calculators will indicate a vertical asymptote when the calculator is set in the DRAWLINE format. They **will not** denote a point discontinuity of this type when calculator is set in either DRAWLINE or DRAWDOT format.

Example 4: Exponential functions

Exponential functions consist of any function of the form a^x, where a is a constant. In business the most common value of "a" is the number e=2.71828..., with some common forms shown below. As you can see, they are all continuous.

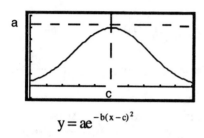

$$y = ae^{-b(x-c)^2}$$

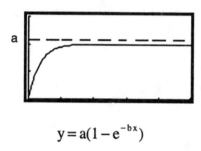

$$y = a(1 - e^{-bx})$$

You will find in the next section that most discontinuities in business type problems are quite obvious. What is not so obvious, and something we will have to look at carefully, are those situations where we have a perfectly nice function until we differentiate it, only to find a discontinuous derivative.

Exercises 5.1

Graph the following functions. Use your graph and the definition of continuous to locate any discontinuities.

1. $y = \dfrac{x^2 - 16}{x + 4}$

2. $y = \dfrac{x - 4}{x^2 - 16}$

3. $y = \dfrac{1}{x - 6}$

4. $y = \dfrac{3x + 5}{2x - 1}$

5. $y = 1000 e^{-.1(x-6)^2}$

6. $y = 2x^2 + 3x - 1$

7. $\dfrac{1}{x^2 + 3}$

8. $y = \dfrac{5}{7} e^{-(x-7)^2}$

Section 5.2 Graphing Piece-Wise Functions

We must now consider some of the more common situations which arise when dealing with functions which arise in the real world. As you will see, many of these have continuous pieces even though the function is not continuous over the entire domain of definition. Of an equally important nature are those functions which are continuous but have discontinuous derivatives. In this section we will learn to deal with many such problems.

For our first example, consider the simple tax table.

Example 1: Jump discontinuity

Any piecewise linear function will be continuous if all left endpoints and corresponding right end points are defined to have the same value. One such example is the way tax tables are often defined. The tax itself, with the possible exclusion of the lowest income groups, is usually a continuous function.

Consider a typical example. Let the tax function T(x) be defined as follows:

$$T(x) = \begin{cases} 0 & 0 < x < 12{,}000 \\ 500 + .23x & 12{,}000 \leq x \leq 25{,}000 \\ .5x - 6250 & 25{,}000 < x < \infty \end{cases}$$

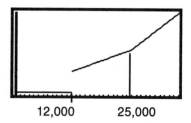

Note: The horizontal segment T(x)=0 is raised slightly to enable us to see that portion of the graph. Normally it would lie on top of the x-axis and be impossible to see.

Looking at the graph of the function you will observe that it has one jump discontinuity. To the left of x=12,000 the value is identically 0, whereas at x=12,000,

T(12000)=500+.23*12000=3,260, which is a jump of 3,260.

Referring to the first two parts of our definition:

1. f(a) exists (you can evaluate function "f" at "a")

   T(12,000)=3,260 therefore part 1 of the definition is satisfied.

2. $\lim_{x \to a} f(x)$ exists (as you approach "a" from either side you get the same value)

In this function, as we approach $12,000 through numbers less than 12,000 T(x) is identically 0, often written as $\lim_{x \to 12000^-} T(x) = 0$; however, if you approach $12,000 through numbers greater than 12,000 T(x) approaches the value $\lim_{x \to 12000^+} T(x) = \$3260$. Since part 2 of our definition fails, the function is discontinuous at x = 12,000.

It is always a conceptual problem for students just beginning to use a graphics calculator to know how to enter an equation like T(x) in the GRAPH editor. To enter piece-wise functions we need the use of inequility symbols and "**and**" (for intersection).

To find the symbols <, ≤, > ≥, press **TEST** (2nd 2).
To find "and" for intersection, press **BASE** (2nd 1), **BOOL** (F4), then **and** (F1).

Here is how we would enter the tax function just discussed in the GRAPH editor:

Set the graph format in **DRAWDOT** (GRAPH, MORE, FORMAT)

y1=(0)(x>0and x<12000)+(500+.23x)(x≥12000 and x≤25000)+(.5x-6250)(x>25000)
or
y1 = (0)(x>0)(x<12000)+(500+.23x)(x≥12000)(x≤25000)+(.5x–6250)(x>25000)

To return to the problem at hand, we are mildly interested in the fact that there is a jump discontinuity at x=$12,000 but our real focus is on what happens to the tax rate at x=$25,000. The function T(x) is continuous here since (referring to our definition of continuous):

1. T($25000) exists, (In fact, T(25000)=$6,250)
2. $\lim_{x \to 25000} T(x)$ exists (approaching from the left $\lim_{x \to 25000^-} T(x)$ you get 6250, and from the right $\lim_{x \to 25000^+} T(x)$ you get closer and closer to 6250, for example, T($25000.01)=$6250.005, and finally,
3. $\lim_{x \to 25000} T(x) = T(25000)$.

Hence, at x=$25,000 the function is continuous, but what about the derivative or tax rate function, T'(x)?
It is often the case in business that the function you are dealing with is continuous at a point but its derivative is not. The tax **rate** is not a continuous function. Recall from our discussion when we first started working with derivatives that the word "**rate**" is often used in place of "**derivative**" or "**slope function**" in the world of

business.

Equivalent statements:

slope function of f(x)

marginal f at a point; for example, if f were a profit function we would say "marginal profit" at x.

f '(x)

derivative of f with respect to x

rate of change of f with respect to x, or, as is more common in business, just "**rate**", such as, if tax is 25% of your income, .25*I, the tax **rate** is .25 which is the slope of the linear equation .25*I.

In our current example, since T(x)=

$$\begin{cases} 0 & 0 < x < 12{,}000 \\ 500 + .23x & 12{,}000 \le x \le 25{,}000 \\ .5x - 6250 & 25{,}000 < x < \infty \end{cases}$$

is just a collection of linear functions we can find the slope at each point easily. Hence, the derivative is T'(x)=

$$\begin{cases} 0 & 0 < x < 12{,}000 \\ .23 & 12{,}000 \le x \le 25{,}000 \\ .5 & 25{,}000 < x < \infty \end{cases}$$

Thus, a picture of the tax **rate** function would look like this.

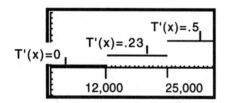
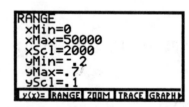

Example 2: More on jump discontinuities

There is always the possibility that both the original tax function AND the tax rate functions will have jump discontinuities as in the following example.

Consider a tax defined on your income x as follows:

$$T(x) = \begin{cases} .25x & 14{,}000 \leq x \leq 35{,}000 \\ .32x & 35{,}000 < x \leq 65{,}000 \end{cases}$$

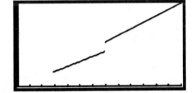 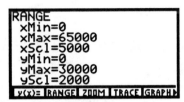

As you can see from the graph and algebraic considerations, there is a jump discontinuity at $35,000. Given no other considerations, your net income N(x) would have a jump discontinuity at the same point, since you could calculate net income as N(x) = x - T(x).

Thus your net income would be

$$N(x) = \begin{cases} x - .25x & 14{,}000 \leq x \leq 35{,}000 \\ x - .32x & 35{,}000 < x \leq 65{,}000 \end{cases}$$

or

$$N(x) = \begin{cases} .75x & 14{,}000 \leq x \leq 35{,}000 \\ .68x & 35{,}000 < x \leq 65{,}000 \end{cases}$$

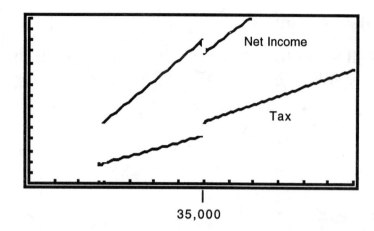
35,000

Getting a raise from $33,000 to $36,000 would yield a decrease in net income:

N(33000)=$24,750

while

N(36000)=$24,480.

How would we enter these functions in our calculator?

Tax: y1=(.25x)(x≥14000) (x≤35000)+(.32x)(x>35000) (x≤65000)

Net Income: y2=(.75x)(x≥14000) (x≤35000)+(.68x)(x>35000)(x≤65000)

Looking at the tax rate we see that it is $T'(x) = \begin{cases} .25 & 14{,}000 \leq x \leq 35{,}000 \\ .32 & 35{,}000 < x \leq 65{,}000 \end{cases}$.

FINDING JUMP DISCONTINUITIES

When functions consist of reasonably simple pieces, curves such as those found in the tax examples 1 and 2, you only need to worry about the possibility of a jump discontinuity at the end points of each interval of definition. For example, in exercise 1,

$$T(x) = \begin{cases} 0 & 0 < x < 12{,}000 \\ 500 + .23x & 12{,}000 \leq x \leq 25{,}000 \\ .5x - 6250 & 25{,}000 < x < \infty \end{cases}$$

calculating T($11,999.99) = 0 and T($12,000) = $500+$2760 = $3260 indicates a jump discontinuity at this point. However, calculating T($25,000) = $500+$5,750 = $6,250 and T($25,000.01) = $6,250.001 indicates that there is no jump discontinuity at these interval end points.

It is not always the case that the function consists of nice straight lines occurring in segments. You can have any combination of curves and straight line segments and some very strange things can happen. Fortunately for you, we are only interested in endpoints matching to the nearest 1¢ and those functions with really weird behavior at end points will be left as curiosities for mathematicians!

One large class of problems that frequently exhibit jump discontinuities are **Economic order quantity** problems (refer to 3.9 in this text). The problem itself requires some thought to understand, so before considering a realistic situation which has jump discontinuities we will review a simplified version.

Example 3: Economic order quantity (continuous)

You decide that over the course of the next year you can sell 50,000 flashlight batteries. Your marginal cost is 85¢ each. However, each order you place costs $40. In addition, your storage costs are 5¢ of the average inventory. How many orders of size "x" should you place to minimize your overall costs?

solution: To resolve this problem you must recognize that it consists of three separate cost functions:

1. **BASE COST OF TOTAL GOODS**
   Overall cost which is 50,000*0.85. No matter what your strategy is, this amount will eventually have to be paid.

2. **INVOICE AND DELIVERY COSTS (somtimes called "set-up costs"**
   Order cost or batch cost (depending upon the situation, sometimes called "set-up costs"). Since each order you place costs $40, you must calculate how many orders you are going to place and multiply that times $40: (# of orders)*(cost of each order). To calculate the number of orders, consider how many orders you would place if "x" were 50,000. Clearly you would place 1 order. If you placed orders of size 25,000 you would place 2 orders. You can see that the number of orders you place is calculated by using the fraction $\left(\frac{50000}{x}\right)$, hence, this part of your cost can be calculated by $\left(\frac{50000}{x}\right)*\$40$.

3. STORAGE COSTS

To calculate storage costs you have to understand the term "average inventory". Average inventory refers to how many items you have in stock on average. For example, if you order x items and start selling them off, at the end of some time period you have none left. Over this time period, what was the average number you had in stock? The answer is $\left(\frac{x}{2}\right)$. Refer to the diagram of example 4 section 3.9. Hence, the cost for storage is $\left(\frac{x}{2}\right)$*$.05.

To calculate the final cost, C(x) of placing several orders of size x, we must add each of the above costs:

1. **BASE COST OF TOTAL GOODS +**
2. **COST PER ORDER FOR INVOICE AND DELIVERY +**
3. **STORAGE COSTS**

C(x)=(base cost of total goods)+(invoice and delivery cost)+(storage cost)

Our example yields the equation:

$$C(x)=50{,}000*0.85+\left(\frac{50000}{x}\right)*\$40+\left(\frac{x}{2}\right)*.05$$

Problems such as this are often referred to as "inventory problems" since the unwary merchant will jump at the chance of making a larger profit by volume purchasing without realizing that it costs money to have stock in inventory and a large inventory can destroy you.

Once you have come up with the correct cost equation, a little calculus comes to the rescue. Differentiating C(x) we get:

$$-\frac{50000}{x^2}*40+\frac{1}{2}(.05)$$

Solving this for x we get x=$\sqrt{80000000}$ ≐ 8944.25. The situation is considerably easier to understand graphically as shown in the following picture:

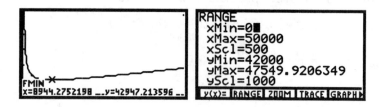

In actual practice most business majors would have trouble working an economic order quantity problem since the cost of the purchase is frequently a discontinuous function. Without specific instruction this is a difficult situation to resolve. The discontinuity is often caused by a volume purchase discount policy that most companies have. For example, purchases of 1-100 cost $9.50 each, purchases from 100-500 cost $8.50 etc. We will see how to resolve this difficulty in the following example.

Example 4: EOQ (discontinuous)

A large retail store can order 27" television sets as follows: For an order between 1 and 19 the price per set is $450 and for 20 or more the price per set is $425. The shipping cost is $90 per order plus $10 for each set in the order (so each set really costs $10 more than the quoted price). Storage costs are $30 times the average inventory. What is the economic order quantity if you know that you must order about a total of 100 sets?

solution: Recall that the solution to such problems is found by calculating:

$$C(x) = (\text{overall cost}) + (\text{batch cost}) + (\text{storage cost})$$

or, in this case,

$$C(x) = (\text{\# sets})*(\text{set cost}) + (\text{\# orders})*(\text{ship cost}) + \text{storage}$$

We can see that we have two situations, one where the best strategy could be to order less than 20 sets, the other when we order more than 20. Let x = # in each order.

1. $x \leq 19$, $C(x) = 100*460 + \left(\frac{100}{x}\right)(90) + 30*\frac{x}{2}$, and

2. $x \geq 20$, $C(x) = 100*435 + \left(\frac{100}{x}\right)(90) + 30*\frac{x}{2}$.

So $C(x) = \begin{cases} 460(100) + \frac{100}{x}(90) + 30\left(\frac{x}{2}\right) & x \leq 19 \\ 435(100) + \frac{100}{x}(90) + 30\left(\frac{x}{2}\right) & x \geq 20 \end{cases}$

As before, we will want to graph this equation to find the minimum value. Graphing this equation yields:

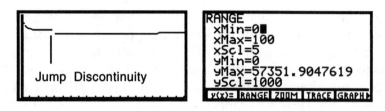

Generally you will need to evaluate the function C(x) at the endpoint(s) of the interval(s). However, if you drew your graph so that you could look fairly closely at C(19) and C(20) you would see immediately that C(19) is not really in the running for least cost. Here are both calculations so that we will have the exact values:

$$C(19) = 100*460 + \frac{100}{19}(90) + 30*\frac{19}{2} = \$46{,}758.60, \text{ and}$$

$$C(20) = 100*435 + \frac{100}{20}(90) + 30*10 = \$44{,}250.$$

Unless we drew our graph REALLY carefully, there is now some question about exactly where the minimum occurs. It is clear that C(20) is close to the best answer, but is it? We may resolve this problem by using the **FMIN** feature to get:

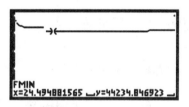

We get the optimal lot size to be 24.4948 and C(24.4948)=$44,234.80. Of course, this optimal solution needs to be corrected to a whole number. A convenient value to pick is x=25 since we need a total of 100 s   You should note here that in general a maximum or minimum could occur at any of the points of discontinuity or any end point, as well as a peak or valley, even though it does not happen in this problem. We will discuss this a little more later on. For now it is sufficient for us to observe that **IF WE GRAPH THE FUNCTION** we can usually spot such occurrences by observation (but not always). In the example we just worked, if we zoomed in around C(20) closely we could see from the graph that the minimum does not occur at C(20), the point of discontinuity, and does occur at the valley we found.

Example 5: Using the "int" function

Suppose your long distance charges were determined as follows: $0.65 connection up to but not including the first minute, $0.60 charge for each additional minute. We assume that calls of duration longer than t.99 minutes would be calculated as t+1 minute calls; 1.991 min. would be billed as a 2 minute call. We will assume the telephone company's clock ticks in hundredths of a minute. Thus, the next tick after .99 is 1.00. Some charges would be as follows:

Time (min.)	.99	.991	1.5	1.991	2.1	2.99
Time+.009	.999	1.000	1.509	2.000	2.109	2.999
charge	$0.65	$1.25	$1.25	$1.85	$1.85	$1.85

To get our calculator to calculate these charges we need a function which only changes values in steps; a **step** function. Such a function is the **int** function (MATH, NUM, int) referred to in mathematics as the "greatest integer" function. This function returns the whole number part of a decimal only. As an example of some values: int(.9)=0, int(.991)=0, int(1.01)=1, int(1.7)=1, int(2.99)=2, etc. We can use this function to calculate the values in the table above. To do this, we can use the function:

$$C(t) = \begin{cases} .65 & t < .99 \\ .65 + .60 * (int(t+.009)) & t \geq .991 \end{cases}$$

Since int(.99)=0, this function reduces to

C(t)=0.65+0.60*(int(t+.009)) for all values of t. The graph is as follows:

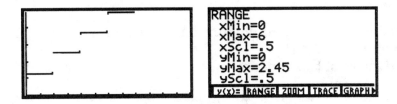

Information needed to work exercise 5.2

1. The ability to graph and carefully adjust both scales.
2. Knowledge of and ability to use the definition of continuity.
3. Knowledge of, and ability to, correctly invoke the "interval function" in your calculator.
4. Basic concepts of a limit.
5. Understanding of how the **int** function works in your calculator.

Exercise 5.2

1. If the tax table were as in example one and your income were close to $25,000, what should your strategy be?

2. If the tax table were as in example 2 and your income were close to $35,000 what should your strategy be?

3. Is it a wise strategy of a government to have jump discontinuities in a tax table?

4. $h(x) = \begin{cases} 2x+3 & x<2 \\ ??? & 2 \leq x \leq 4 \\ 3x+2 & 4<x \end{cases}$.

    a. Find a linear expression for ??? to make h(x) continuous.
    b. Graph the equation h(x) with this linear equation inserted.

5. A laborer is paid $8.50 an hour and time and a half for overtime (over eight hours per day).

    a. Write the function for the total amount paid to the worker per day for t hours work and graph it for $0 \leq t \leq 16$.
    b. Write the derivative of this function for t hours work and graph it for $0 \leq t \leq 16$.
    c. Is the function continuous?
    d. Is the derivative continuous? Explain your answers on the graphs you draw. Use the term "pay rate" instead of derivative in your explanation.

6. A village charges residents for water in the following way:
   $20 flat rate and 20¢ for every 100 gallons beyond 2000 gallons.

   a. Write the function for the total cost of x gallons of water and graph it.
   b. Is the function continuous?
   c. Is the derivative of the function continuous?
   d. If the wording were changed to read "and 20¢ for every 100 gallons or fraction thereof beyond 2000 gallons", how would this affect your equations and write the new equations down.

7. $f(x) = \begin{cases} x^3 & x < -1 \\ x^2 & -1 \leq x \leq 1 \\ x^3 & x > 1 \end{cases}$. Is f(x) continuous at 1? at -1? Is f '(x) continuous at 1? at -1?

8. An electric company has a minimum charge of $8.50 per month for any usage up to 1000 kilowatt hours (kwh). For usage from 1000 up to and including 10,000 kwh, the rate is $8.50 plus $.05 per kilowatt hour for all kwh over 1,000. For usage above 10,000 kwh it is $458.50 plus a charge of $.045 per kilowatt hour for all usage over 10,000 kwh.

   a. Write this function in standard "human" notation (you know, with the brackets like your math teacher would).
   b. Write the equation in calculator notation using the special interval function.
   c. Graph
   d. Give the slope function or derivative of this function in each interval in regular old human notation.
   e. Write this slope function as a single function using the special calculator notation and graph it.

9. A five year C.D. has no penalty for early withdrawal. To encourage people not to remove their money early, the bank offers an increasing interest rate as shown:

   5% for the first year, 8% for the second and third years, 11% the fourth year, and 14% the final year.
   a. Write the function for the total amount received upon withdrawing a deposit t years after a $3000 deposit. Assume interest is continuous.

   Hint: The first year's equation is easy: 3000e$^{.05t}$, 0<t<1. However for years 2 and 3 the equation will be:

   $$3000e^{.05(1)} * e^{.08(t-1)}, \quad 1 \leq t < 3.$$

   b. Graph this function.

10. Suppose the C.D. in number 9 carries a penalty of $1000, deducted if the deposit is withdrawn within the first nine months. Write the function for the total amount received upon withdrawing a deposit t years after a $3000 deposit and graph it.

Hint: To work this problem you may use all your results from #9 except for the first year. The first year must be split into two equations, one for $0 < t < \frac{9}{12}$, and another for $\frac{9}{12} \leq t < 1$.

11. A telecommunications company offers long distance service at the following rate: 35¢ for connection and charge for the first <u>full</u> minute and 12¢ for each minute (or fraction of a minute) after the first. We will assume the telephone company's clock ticks in hundredths of a minute. Thus, the next tick after .99 is 1.00. Unlike example 4, charges increase only after the full minute tick, eg. the charge for 1.00 minutes is 35¢; the charge for 1.01 minutes is 47¢. Using the **int** for the "greatest integer" function,

   a. Write the equation(s) for the cost of a t minute call.
   b. Is the function continuous at t=1 minute?
   c. Calculate charges for calls of length .5, 1, 1.01, 1.99, 2, 2.01, and 3.01 minutes.
   d. Graph this function for x = 0 to 10.

12. Rework example 3 from the text but instead of assuming a flat wholesale cost of 85¢ each for all 50,000 batteries, assume that a 10% discount is given for orders of 10,000 or greater.

   a. Write down these new equations
   b. Graph the new equations.
   c. What is the optimum economic order quantity now? After calculating this quantity, label it clearly on your graph.

13. Rework exercise 12 with the 10% discount for orders in excess of 10,000, but assume that order costs increase to $150 per order for orders over 10,000 items.

   a. Write down these new equations
   b. Graph the new equations.
   c. What is the economic order quantity now? After calculating this quantity, label it clearly on your graph.

14. A store has the opportunity to buy an item in volume at a significant saving. As an example, the item costs $35 each for up to 500 items. 500-2000 items cost $30 each. The store estimates that it can sell about 750 of these items in the next six months. Since repeated handling is involved, storage costs are rather high at $5 per item on the average inventory and sales are linear. Shipping costs are $75 per order plus 95¢ for each item in the order.

    Compare the strategies of:

    a. ordering x units at a time, 0<x<500, and reordering as needed; ie., what is the economic order quantity C_1 when the order size x is less than 500 and what is the cost at that quantity?
    b. ordering 750 at once; ie., what is the cost C_2 of ordering all 750 at one time?
    c. ordering 500 in the first order and 250 in a second order.

15. I noticed that our last telephone bill has a new rate: $17.32 for basic service and a flat 10¢ a minute for long distance calls at night. Assuming I only make long distance calls at the new night rate, write out the telephone bill function. Graph this function for $0 \leq t \leq 10$. Recall that int(.35)=0, and int(1.01)=1.

16. A golf course charges a membership fee of $500 per year plus $7 for every nine holes (or fraction of) that you play per day. Use the **sum seq** feature and the **int** feature of the MATH editor to :
    a. Write the daily cost function for golf using the int function. (Hint: Let g be the number of holes played and do something clever with the fraction g/9).
    b. Write the yearly cost of playing n times.

17. Ice scrapers cost $1.20 each with a batch cost of $200 per order, regardless of size. However, there is no storage costs for orders of less than or equal to 2000, but for orders larger than that the storage costs become 10% ($0.12) of the average inventory costs. If you predict sales to be 50,000 during the course of the next year,

    a. Write the equation for the cost of placing orders of size 2000 or less. Calculate the cost of placing enough orders of size 2000 to fill your needs.
    b. Write the equation for placing orders of more than size 2,000. Draw a graph of this equation and calculate the size order that will yield minimum cost and what is that minimum cost.
    c. What size order is best?

If you have worked a fairly good selection of problems up to this point, you are ready for an actual situation. Calculus (or at least the ideas contained therein) can be used to help you find the best solution. The word "best" must now be interpreted in light of the audience. On the one hand, your math professor will be delighted if you find the minimal cost when ordering 50,000 items occurs at an order size of 17,456.89 or some such answer. You, as a business student, must recognize that this is an impossible situation. 17,456.89 is not a divisor of 50,000 and one would never place such orders. More useless math? (I love the term Japanese students have coined for this "juuken sugaku" - "test-taking mathematics").

We now make the connection between the calculus answer and the real world. Since we know that the ideal answer will occur around 17,500 we consider a few combinations. How about two orders of 20,000 and one of 10,000? Or, two orders of 17,000 and one of 16,000? Calculus has given us a shopping center guide: **YOU ARE HERE** and now you must simply look about you to see exactly what needs to be done. In actual situations, you will find only slightly different answers when you make slightly different beginning choices.

18. A product sells for $1 in orders of size x less than 5000, but for .75 in orders of size x≥5000. There are no storage costs for orders of size less than 5000, but orders of 5000 or greater incur storage costs of $.08 each on the average inventory. All orders cost $250 to place and process. It is known that about 50,000 will be needed.

   a. Write down the cost equation.
   b. Find the exact optimal order size and the cost at that point as given by your equation.
   c. Interpret your answer to part b given that you cannot place fractional orders.
      1. Exactly how many orders should you place?
      2. What is the size of each of these orders?
      3. What is the total cost of placing orders in this fashion?

19. A product sells for $1 in orders of size x less than 5000, but for .75 in orders of size x≥5000. There are no storage costs for orders of size less than 5000, but orders of 5000 or greater incur storage costs of $.10 each on the average inventory. All orders cost $200 to place and process. It is known that 50,000 will be needed.

   a. Write down the cost equation for your math program. If you choose not to use the interval function, make sure you clearly indicate which equation goes with which range of orders.
   b. Find the exact optimal order size and the cost at that point as given by your equation.

c. Interpret your answer to part b given that you cannot place fractional orders.
   1. Exactly how many orders should you place?
   2. What is the size of each of these orders?
   3. What is the total cost of placing orders in this fashion?

20. Ice scrapers cost $1.20 for orders of size 10,000 or less and $0.65 for orders greater than that. Order cost is $200 per order, regardless of size. The storage costs are 10% ($0.12) of the average inventory costs for orders up to size 10,000 and 16% on the average inventory for orders larger than that. If you predict sales to be 50,000 during the course of the next year,

    a. Write the cost equation for placing orders.
    b. Draw a graph of this equation, calculate the exact size order that will yield minimum cost, calculate that minimum cost and label that point on your graph.
    c. Now give a good order combination (just to the nearest 1,000's) which will give a good minimum cost and total 50,000 items (like 2-20,000 and 1-10,000) and give the total cost for those orders.

Major homework assignment:

Roby Triplett, our bookstore manager, passes along to you the following problem. This problem is typical of the many he faces every day in ordering materials for the bookstore and attempting to minimize the cost. Any savings the bookstore realizes are passed along to you, the student. As a former ASU student he constantly strives to minimize his costs to provide savings for the student body.

He has never viewed this as a calculus problem, not because it isn't, or that he was a poor math student, which he wasn't, but because of the way calculus was presented to him. As I have stated to you many times, businessmen rarely get continuous functions, or even easily recognized functions, to deal with. Your task is to solve this problem using your knowledge and to present your solution to him in a form which he can read and understand. Include graphs and all pertinent explanations. He has not had any math courses in many (many, many) years.

THE PROBLEM

The ASU bookstore purchases composition and note books (referred to as "wirebounds" in the trade). In a normal year they will retail about $33,000 worth. They sell about 80% of those in the fall semester and about 20% in the spring semester. On orders of $0 to $2,000 they get a 40% discount off retail price. On orders of $2,000 to $4,000 they get a 50% discount. Orders over $4,000 yield a 50-10 discount. For example, a $10,000 retail order would cost:

$10,000-(50% of 10,000)=5,000; 5,000-(10% of 5,000)=4500;

so the $10,000 order would cost $4,500. On orders from $0 to $2,000 they have no significant storage and handling costs since all the goods can be put directly in the store. The cost of placing an order is about $200 for orders up to $2,000 about $400 for orders from $2,000 to $4,000 and about $500 for orders larger than that.

First, disregard the 80% - 20% split in sales peculiar to wirebounds and find their economic order quantities for wirebounds given that

a. If they order over $2,000 then the excess has to be put in the warehouse incurring additional costs of about 12% on the inventory placed in the warehouse. If they order over $4,000 worth an additional cost of 3% of the inventory placed in the warehouse over $4000 must be added to the storage cost.

b. If they order over $2,000 then the excess has to be put in the warehouse incurring additional costs of about 12% on the average inventory in the warehouse. If they order over $4,000 worth an additional cost of 3% of the average inventory over $4000 must be added to the storage cost, and

c. Finally, write a paragraph explaining what their strategy must be given the 80% - 20% split in sales. You might also want to invoke the fact that there is about a 6 week delay between the placing of an order and the receipt of the goods.

Section 5.3 Elasticity

It is the nature of human beings to want to know what is going to happen in the future. In the past, we have spent money on people with crystal balls, dark rooms with vibrating tables, and the like. In business we call such people economists. It is their job to try to peer into the future and say what is going to happen, what is happening, and has happened. They do this with varying degrees of success, although often even their successes are not impressive. For example, we know that if interest rates drop too much, inflation will set in. What is "too much"? Too much is whenever inflation sets in; a not entirely acceptable answer.

One relatively simple calculation with some value in peering into the future is called the "elasticity of demand". This is a fairly simple yet sophisticated number which can be calculated easily that reflects market reaction to any change in price.

To introduce this concept, let's start with an unlikely example. Suppose I tell you that I have two sets of animals and I try a diet on them designed to effect weight gain. After a period of time, we take some measurements to determine the effectiveness of the diet. With group E we discover that average weight gain during a time period t is 200 lbs., whereas with group M we discover that the average weight gain is 3 ounces. Given this information you might conclude that the diet is more effective on group E than on M.

The difficulty with such a superficial analysis becomes apparent when I tell you that E stands for elephants and M stands for mice. To make any sense out of the results, we must somehow come up with a comparison standard that eliminates the differences in the animals and the weight measurements. One such measurement can be derived as follows:

First, we take the change in weight of each animal and divide it by the weight. For example, a mouse might weigh in at 8 ounces. The change in weight for the mice M is 3 ounces. We have been writing the term "change in" as Δ. Thus, the change in the mice's weight would be written ΔM. Using this notation we form the ratio $\frac{\Delta M}{M} = \frac{3 \text{ ounces}}{8 \text{ ounces}}$ and get the **number** $\frac{3}{8}$. It is crucial to notice here that units have been eliminated by the division; ie., we started with units in ounces and ended with a pure number. Now do the same with the elephants, assuming that an elephant weighs in at about 3500 lbs. Putting the change in elephants weight over the weight we get $\frac{\Delta E}{E} = \frac{200 \text{ lbs.}}{3500 \text{ lbs.}}$. Canceling units yields $\frac{200}{3500}$. We can now make a meaningful comparison of the dietary effects between animals: $\dfrac{\frac{\Delta M}{M}}{\frac{\Delta E}{E}} = \dfrac{\frac{3}{8}}{\frac{200}{3500}}$.

Reducing this fraction we get $\frac{3}{8} \times \frac{3500}{200} = 6.5625$. In other words, the diet is about six and a half times as effective on the mice as it is on the elephants.

In business, we often wish to compare what happens to one function when something else which affects it changes in value. An example of this is to consider what happens to the demand price when the quantity is changed. In order to accomplish this we must find some way to compare dissimilar things, like prices and quantities. This is where our idea in the mouse vs. elephant problem will pay off. What we wish to know (our crystal ball) is what effect on sales, revenue, and profit will a change in the price charged cause. Clearly one of the most important questions for us to answer (ahead of time) is:

What is the most we can charge for our product before consumer resistance and competition set in to cause a drop in revenue?

Note that revenue can actually be increased if the INCREASE IN PRICE MORE THAN OFFSETS THE DROP IN SALES. This is where the **elasticity of demand** shows its value.

The elasticity of demand (or just **elasticity**) is a measure of the relationship between the demand price and sales. Given a demand price p, what happens to the sales when this price is raised or lowered?

Example 1: Sheet rock

Since sheet rock has few suppliers, is expensive to import due to weight considerations and is in constant if fluctuating demand, producers can keep a close watch on the elasticity of demand. A recent price change produced the following results:

	November 91	March 92
Price	$10.00	$12.00
Quantity	1.75 units	1.31 units

Was this price increase a wise choice? Was it wise to raise the price during a period when there is a seasonal drop in purchases? What should be done next? To answer these questions we need to have some way to compare the relationship between the change in price and the change in quantity sold - the mice and the elephant problem. We will solve it in the same fashion. First, we can eliminate units by forming the ratio of change in price to price $\frac{\Delta p}{p}$ and change in quantity to quantity $\frac{\Delta q}{q}$. Since both Δp and p are in dollars, the ratio has no units. The same is true of Δq and q. Thus, we may now compare these numbers in a second ratio $\dfrac{\frac{\Delta q}{q}}{\frac{\Delta p}{p}}$ to see their relative effects. This leaves us with only one remaining problem. This number will (nearly) always be negative and SOME PEOPLE make lots of mistakes when dealing with negative numbers.

To see why the fraction $\dfrac{\frac{\Delta q}{q}}{\frac{\Delta p}{p}}$ is negative, consider the accompanying diagram.

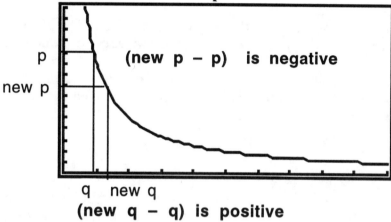

Under normal circumstances a consumer demand equation is a decreasing function and consequently, a decrease in price (a **NEGATIVE** increase) will be accompanied by an increase in sales. In this figure, an increase in price is accompanied by a decrease in sales, so $\dfrac{\Delta q}{q}$ is positive but $\dfrac{\Delta p}{p}$ is negative. The fraction $\dfrac{\frac{\Delta q}{q}}{\frac{\Delta p}{p}}$ is negative.

Considering the price increase shown in the following diagram. The elasticity of demand ratio is still negative since a price increase will be accompanied by a corresponding decrease in sales.

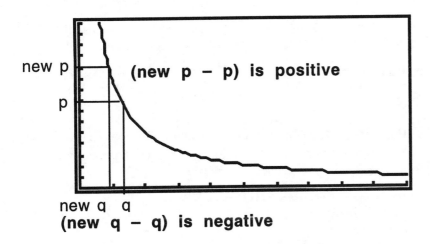

Hence, $\frac{\Delta q}{q}$ is negative but $\frac{\Delta p}{p}$ is positive. Thus, the fraction $\frac{\frac{\Delta q}{q}}{\frac{\Delta p}{p}}$ is again negative.

To avoid always working with a negative number, we put a minus sign in front of our ratio.

$$\text{This fraction,} \quad -\frac{\frac{\Delta q}{q}}{\frac{\Delta p}{p}}$$

is called the **elasticity of demand.** We will be particularly interested to see if any decrease in price is more than offset by a large increase in sales since this would yield an overall increase in revenue. Under some circumstances we can raise the price and there is no correspondingly large drop in sales, again yielding a revenue increase.

It is this increase - decrease relationship that we wish to study.

There is a problem with this elasticity of demand formula which we must correct before proceeding. Consider the following two conditions:

1. The numerical value of the elasticity should be the same whether we consider the problem as a price increase or a price decrease problem. However, we viewed the denominator of each fraction $\frac{\Delta q}{q}$ and $C(q) = \frac{1}{75}q^2 + \frac{q}{6} + 75$ as the **starting** value. Since the numerators will have the same numerical value regardless of the starting point, changing the values of the denominators will affect the value of the fraction.

2. Ultimately we will be most interested in what affect a price change will have on the revenue. In our delta notation, what affect does a change in price have on the revenue? This can be calculated by calculating the value of $\frac{\Delta R}{\Delta p}$. As we shall see, the **elasticity** will have a direct bearing on this ratio.

Neither of these concerns will be met with our current equation for elasticity. First consider the situation where we view the problem first as a price increase problem, then as a price decrease problem using our sheet rock example:

	November 91	March 92
Price	$10.00	$12.00
Quantity	1.75 units	1.31 units

First, consider this as a price increase problem.

If we had started with $10 and raised the price to $12, $\frac{\Delta p}{p} = \frac{10-12}{10} = -\$2/\$10$ =-0.2 and $\frac{\Delta q}{q} = \frac{1.75 - 1.31}{1.75} = .25$ so $-\frac{\frac{\Delta q}{q}}{\frac{\Delta p}{p}} = 1.257$. On the other hand, if we had started with $12 and viewed the problem as a price decrease problem we would get $\frac{\Delta p}{p} = \frac{12-10}{12} = \$2/\$12 = 1/6$ and $\frac{\Delta q}{q} = \frac{1.31 - 1.75}{1.31} = -0.336$ so $-\frac{\frac{\Delta q}{q}}{\frac{\Delta p}{p}} = 2.015 \neq 1.257$. We can see that the problem with dividing by the initial price is that the price elasticity is different depending on the direction of the price change. The elasticity coefficient, if it is to be useful, should be the same for a given range of prices irrespective of whether the price change being considered is a price increase or a price decrease. There should not be one elasticity for price decreases and another one for price increases.

This problem can be resolved by using the averages $\frac{(\text{old q} + \text{new q})}{2}$ in place of q in the denominator of $\frac{\Delta q}{q}$ and $\frac{(\text{old p} + \text{new p})}{2}$ in place of p in the denominator of $\frac{\Delta p}{p}$. To see this, look at the same calculations again.

With $10 and $12 as the two prices, the average price is ($10+$12)/2=$11; thus $\frac{\Delta p}{\frac{\text{old p + new p}}{2}} = \frac{(10-12)}{11} = .18$. With 1.31 and 1.75 as the quantities, (1.31 + 1.75)/2= 1.53 so $\frac{\Delta q}{\frac{(\text{old q + new q})}{2}} = \frac{(1.75 - 1.31)}{1.53} = 0.288$. Thus the value of the elasticity is .288/.18=1.58 (approximately).

This calculation uses the **midpoint** of each interval. As you can easily see, by using these midpoints it doesn't matter if we view the problem as an increasing or decreasing cost problem. The midpoints formula uses the average price and average quantity to calculate the price elasticity of demand. It yields the same elasticity coefficient for an increase from a lower price to a higher price as for a decrease from a higher price to a lower price.

Hence, the corrected formula will be:

Midpoints formula for Elasticity of Demand: $E = -\dfrac{\dfrac{\Delta q}{\left(\dfrac{\text{old q + new q}}{2}\right)}}{\dfrac{\Delta p}{\left(\dfrac{\text{old p + new p}}{2}\right)}}$

As we shall see next, this midpoints formula corrects both problem 1 and 2 stated above. It is absolutely necessary to make the ratio $\frac{\Delta R}{\Delta p}$ correctly reflect the relationship between revenue, price change, and elasticity.

Returning to our sheet rock problem again, recall that the elasticity was 1.58. We now come to one of the most difficult points in working with the elasticity of demand. What does a final result of 1.58 mean? Since we are dealing with a ratio with the numerator a measure of change in quantity sold and the denominator a measure of change in price charged, it is apparent that the change in quantity was about 1.58 times as severe as the change in price. This does not sound good. If you will think about it for a moment, you will realize that this kind of disproportionate drop in sales could severely affect the revenue. What, exactly is the relationship between these factors?

Recall that Revenue = (price) x (quantity). Looking at our data again we find that in November, demand price was $10.00 and quantity was 1.75 units,

	November 91	March 92
Price	$10.00	$12.00
Quantity	1.75 units	1.31 units

so

November revenue is 17.50 unit dollars.

In March, price was $12.00 and quantity was 1.31 units, so

March revenue was 15.72 unit dollars

which is indeed a drop in revenue. Thus, **raising the price did not raise the revenue** as we hoped. Hence, our suspicions are well founded. The elasticity of demand indicates that **the drop in sales more than canceled any benefit derived from raising the price.**

We now ask the following question. What is a "good" elasticity number - good in the sense that a raise in price will either produce an increase in revenue, or, at the very least, no drop? It should immediately be apparent to you that an elasticity of 1 would indicate the same relative change in both price and quantity. Such an elasticity should indicate that a relatively small change in price will have little or no change on revenue.

In addition, if the elasticity is less than 1 it would mean that the numerator $\dfrac{\Delta q}{\frac{(\text{old } q + \text{new } q)}{2}}$ changed relatively less than the denominator $\dfrac{\Delta p}{\frac{\text{old } p + \text{new } p}{2}}$. This would indicate less change in quantity than price. Would this indicate an increase in revenue?

To see the answer to this question we need to do a little calculation. First, the question we are asking needs to be very clear in your mind.

What effect does a change in price have on the revenue?

In terms of our Δ notation, we wish to know the value of the fraction $\dfrac{\Delta R}{\Delta p}$ under differing circumstances. As we shall now see, only the midpoints formula will give the desired result.

The relationship between $\frac{\Delta R}{\Delta p}$ and the

elasticity of demand $E = -\dfrac{\dfrac{\Delta q}{\left(\dfrac{\text{old } q + \text{new } q}{2}\right)}}{\dfrac{\Delta p}{\left(\dfrac{\text{old } p + \text{new } p}{2}\right)}}$

Consider both p (demand price) and revenue R as functions of the quantity q. Using standard function notation we express both of these functions as R(q) and p(q). Hence, the revenue function R(q) can also be expressed as

$$\text{Revenue} = \text{price} * \text{quantity}$$

or

$$R(q) = p(q) * q$$

We now need to express the fraction $\frac{\Delta R}{\Delta p}$ using this notation. If we select q_1 for our original or starting quantity and q_2 for our changed quantity ΔR is $R(q_2)-R(q_1)$ and $\Delta p = p(q_2)-p(q_1)$. Hence, the fraction becomes $\dfrac{\Delta R}{\Delta p} = \dfrac{R(q_2)-R(q_1)}{p(q_2)-p(q_1)}$.

Good news - bad news

Before doing our calculations consider the following two situations. Since we have good news and bad news, let's get the bad news out of the way first.

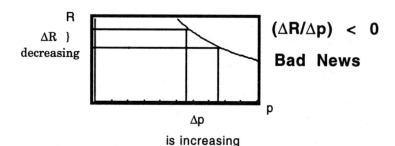

$(\Delta R/\Delta p) < 0$

Bad News

Δp is increasing

In this first case, as the price increases the revenue drops. Since price increases, $\Delta p = p(q_2) - p(q_1)$ is positive. Likewise, since revenue decreases, $\Delta R = R(q_2) - R(q_1)$ will be negative so our fraction $\dfrac{\Delta R}{\Delta p} = \dfrac{-}{+} < 0$.

On the other hand we could view this as a price decrease with an accompanying quantity increase. In this case, $\dfrac{\Delta R}{\Delta p} = \dfrac{+}{-} < 0$, the same result as before.

On the good news side, there is always the possibility that raising the price will cause revenues to increase.

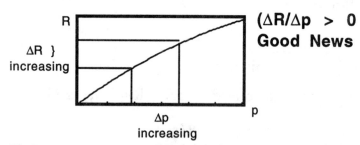

$(\Delta R/\Delta p) > 0$

Good News

Δp increasing

If this should happen then both $\Delta p = p(q_2) - p(q_1)$ and $\Delta R = R(q_2) - R(q_1)$ are positive and hence the fraction $\dfrac{\Delta R}{\Delta p}$ will be positive, $\dfrac{\Delta R}{\Delta p} = \dfrac{+}{+} > 0$.

On the other hand, this could be viewed as a situation where lowering the price causes a drop in the revenue, $\dfrac{\Delta R}{\Delta p} = \dfrac{-}{-} > 0$. The fact to note here is that the ratio $\dfrac{\Delta R}{\Delta p} > 0$ in both cases.

Question: Is there anything in the numerical value of the elasticity which will quickly tell us which of the two situations for $\dfrac{\Delta R}{\Delta p}$ we have?

Answer:
The answer is yes. Since we couldn't find this work anywhere else, we include the following algebraic exercise. If you are curious, you should read it - the algebra is not

difficult. However, feel free to take the result on faith if you wish and skip to "**END OF ANSWER**".

Since R=pq,

$$(\text{change in R}) = \Delta R = \Delta(pq) = (\text{change in pq}).$$

Let R_1 and p_1 be the values of the revenue and price at quantity level q_1. Let R_2 and p_2 be their values at quantity level q_2. Thus, $\Delta R=[R_2-R_1]$ and $\Delta p=p_2-p_1$. Also, since $\Delta(pq)=pq$ evaluated at q_2 minus pq evaluated at q_1, we have $\Delta(pq)=p_2q_2-p_1q_1$. So, $\frac{\Delta R}{\Delta p}=\frac{p_2q_2-p_1q_1}{\Delta p}$. We must now pull a mathematical trick. The numerator of this fraction, $p_2q_2-p_1q_1$, is not in a usable form for what we need to do next. Its form can be changed by adding $p_2q_1-p_2q_1$ (which is 0) in the middle of the equation:

$$p_2q_2-p_1q_1=p_2q_2+\{p_2q_1-p_2q_1\}-p_1q_1$$

Now factor as follows:

$$p_2q_2+p_2q_1-p_2q_1-p_1q_1=p_2[q_2-q_1]+q_1[p_2-p_1].$$

The right hand side can be rewritten using our delta notation as

$$p_2[q_2-q_1]+q_1[p_2-p_1]=p_2*\Delta q+q_1*\Delta p$$

Hence, $\frac{\Delta R}{\Delta p}=\frac{p_2\Delta q+q_1\Delta p}{\Delta p}$. Putting the two terms in the numerator over the denominator and canceling we get

$$\frac{\Delta R}{\Delta p}=p_2\frac{\Delta q}{\Delta p}+q_1$$

Recall now that we were looking at the bad news-good news scenarios. In the case of bad news, $\frac{\Delta R}{\Delta p} < 0$. Hence, $p_2\frac{\Delta q}{\Delta p}+q_1 < 0$. This is not quite what we need since we are trying to show the connection between elasticity and change in revenue.

Consider again the equation $p_2q_2-p_1q_1$ and add $p_1q_2-p_1q_2$ (which is 0) instead of $p_2q_1-p_2q_1$ which we added last time. We now get:

$$p_2q_2 - p_1q_1 = p_2q_2 + \{p_2q_1 - p_2q_1\} - p_1q_1$$

and factoring as we did before, we get

$$p_2q_2 + \{p_2q_1 - p_2q_1\} - p_1q_1 = p_1(q_2 - q_1) + q_2(p_2 - p_1)$$

which can be rewritten as:

$$p_1(q_2 - q_1) + q_2(p_2 - p_1) = p_1\Delta q + q_2\Delta p.$$

so our original expression can also be rewritten as

$$\frac{\Delta R}{\Delta p} = p_1 \frac{\Delta q}{\Delta p} + q_2$$

Thus we have two expression for the same quantity:

$$\frac{\Delta R}{\Delta p} = p_2 \frac{\Delta q}{\Delta p} + q_1 \quad \text{and} \quad \frac{\Delta R}{\Delta p} = p_1 \frac{\Delta q}{\Delta p} + q_2$$

Now consider again the bad news-good news scenarios. In the case of bad news, $\frac{\Delta R}{\Delta p} < 0$. Hence, using both expressions $p_1 \frac{\Delta q}{\Delta p} + q_2 < 0$ and $p_2 \frac{\Delta q}{\Delta p} + q_1 < 0$, we can add the two left hand sides getting $p_1 \frac{\Delta q}{\Delta p} + q_2 + p_2 \frac{\Delta q}{\Delta p} + q_1 < 0$. Combining and factoring we get $(p_1 + p_2)\frac{\Delta q}{\Delta p} + (q_2 + q_1) < 0$ so $(q_2 + q_1) < -(p_1 + p_2)\frac{\Delta q}{\Delta p}$ and $1 < -\frac{(p_1 + p_2)\Delta q}{(q_1 + q_2)\Delta p}$. Rewriting this last expression we see that it is nothing but the elasticity and states that when revenue decreases with a price increase so that $\frac{\Delta R}{\Delta p} < 0$, the elasticity is greater than 1. Since each of the above steps is merely an algebraic manipulation, we could have reversed our reasoning, starting with elasticity>1 and arriving at $\frac{\Delta R}{\Delta p} < 0$. Thus we see that when the elasticity is greater than one, revenue decreases with any price increase:

(END OF ANSWER)

Elasticity > 1, Revenue vs. Price

$$E = -\frac{\frac{\Delta q}{(q_1+q_2)}}{\frac{\Delta p}{(p_1+p_2)}} > 1 \text{ if and only if } \frac{\Delta R}{\Delta p} < 0$$

The term "elastic" is attached to this situation since an increase in price causes a decrease in revenue and a decrease in price causes an increase in revenue: price and revenue move in opposite directions. You can remember this by thinking of revenue and price being attached by a rubber band. Since the connection is elastic, they can move in opposite directions.

Returning to our graph of Revenue vs. price, we see this relationship in the graph, elasticity

$$E > 1 \text{ if and only if } \frac{\Delta R}{\Delta p} < 0$$

That is, if elasticity E >1, price and revenue move in opposite directions.

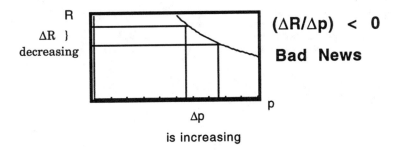

$$\frac{\Delta R}{\Delta p} = \frac{-}{+} \text{ or } \frac{+}{-} < 0$$

When $\frac{\Delta R}{\Delta p} > 0$, the same reasoning yields:

Elasticity < 1. Revenue vs. Price

$$E = -\frac{\frac{\Delta q}{(q_1+q_2)}}{\frac{\Delta p}{(p_1+p_2)}} < 1 \text{ if and only if } \frac{\Delta R}{\Delta p} > 0$$

That is, when elasticity is less than one, the revenue either decreases with a price decrease, or increases with a price increase. The term "inelastic" is attached to this phenomenon since both price and revenue move in the same direction.

Returning to our graph of Revenue vs. price, we see this relationship in the graph. Elasticity

$$E < 1 \text{ if and only if } \frac{\Delta R}{\Delta p} > 0$$

That is, if elasticity E<1, price and revenue move in the same directions.

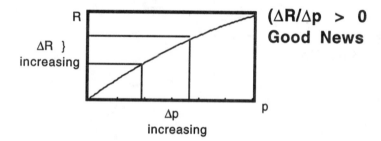

If $\frac{\Delta R}{\Delta p} > 0$, then

$\frac{\Delta R}{\Delta p} = \frac{+}{+} =$ (both increase) or $= \frac{\Delta R}{\Delta p} = \frac{-}{-}$ (both decrease).

Example 2: Short and long term elasticities

In your economics course you will discover that elasticities for many products and services are well known. These elasticities are usually organized into two effect groups: short term and long term. Consider the following table:

	short term	long term
jewelry	.4	.7
electricity	.13	1.9

In the short term, what effect does an increase in price have on the revenues for jewelry? For electricity? What about the long term effect?

solution:

Jewelry:

Since $1 > -\dfrac{\frac{\Delta q}{(q_1+q_2)}}{\frac{\Delta p}{(p_1+p_2)}}$ if and only if $\dfrac{\Delta R}{\Delta p} > 0$ and the elasticity E<1, in both short and long term, $\dfrac{\Delta R}{\Delta p}$ being positive means that either both revenue and price increase together or decrease together (inelastic); raising the price will raise the revenue, lowering the price will lower the revenue. If that's the case, then why do we always see jewelry "on sale?"? Recalling some of the litigation of the government vs. merchants having "sales" will help you determine the answer to this question.

Electricity:

In the short term, elasticity being as small as .13 indicates that a rise in prices will cause a (fairly significant) increase in revenues. Interpreting this result in terms of your own household, it means that your parents are not going to yell at you until after the first electric bill comes in. At which point "turn off the lights" will become the household cry.

However, in the long term, the condition E>1 if and only if $\dfrac{\Delta R}{\Delta p} < 0$ indicates that after some period of time the price and revenues will head in opposite directions (elastic demand) because consumers will reduce their overall consumption of electricity thus causing a drop in revenues.

Example 3: Elasticity in pricing strategies

A company has gathered the following data on sales for one item in the store.

price	3.59	3.39	3.89	3.19	4.59
quantity	5200	5600	4800	5700	1200

If we consider only revenue, what pricing strategy would seem to be the best?

solution:

First we **rearrange the data in ascending order by price**, then calculate both revenue and elasticity.

Note: Enter the elasticity formula in the SOLVER. Remember to enter it from the HOME screen to store it in the SOLVER.

```
ELAS=-((NQ-OQ)/(NQ+OQ
))/((NP-OP)/(NP+OP))
                Done
```

To fill in the table for elasticity, use the SOLVER as follows:

```
exP=-((NQ-OQ)/(NQ+OQ...
■exP=.29115044247787
NQ=5600
OQ=5700
NP=3.39
OP=3.19
 bound=(-1E99,1E99)
GRAPH RANGE ZOOM TRACE SOLVE
```

```
exP=-((NQ-OQ)/(NQ+OQ...
■exP=1.2925925925926■
NQ=5200
OQ=5600
NP=3.59
OP=3.39
 bound=(-1E99,1E99)
GRAPH RANGE ZOOM TRACE SOLVE
```

price	3.19	3.39	3.59	3.89	4.59
quantity	5700	5600	5200	4800	1200
revenue	18,183	18,984	18,668	18,672	5,508.
elasticity		.29	1.29	.997	7.27

You can notice several crucial facts from looking at the numbers. First, elasticity (.29) is very low as price changes from $3.19 to $3.39 (very inelastic). This would indicate that a rise in price should cause a rise in revenues, which it does. Between $3.39 and $3.59 the elasticity is 1.29, indicating a price rise will cause a drop in revenues. What actually happens with the price rise from $3.59 to $3.89 is a slight increase in revenue. As the price rises from $3.89 to $4.59 we find an elasticity of 7.27, very elastic, with the expected drop in revenue. How are we to interpret these results; what is the best pricing strategy?

If you consider the actual marketplace, nothing operates in a vacuum. As your prices vary, the competition tries to outdo you in one way or another. Hence, your price rise might have caused another merchant to make a significant price reduction attracting a significant amount of your business. So is all lost? Not really. Consider the graphs of revenue vs. elasticity.

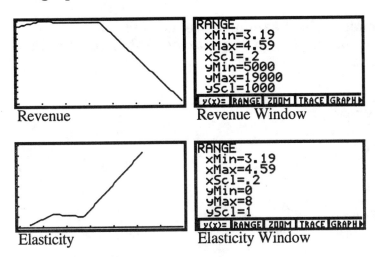

Revenue   Revenue Window

Elasticity   Elasticity Window

Looking at this data graphically, our strategy should be clear. At a price of around $3.59 our elasticity hovers around unity. This means that given whatever market pressures there are, we will be near the maximum revenue level when we are around this price.

Example 4: Using elasticity on ratio $\dfrac{\Delta R}{\Delta p}$

Finally, let us consider a rough interpretation of the meaning of the fraction $\dfrac{\Delta R}{\Delta p}$. Suppose the numerical value of $\dfrac{\Delta R}{\Delta p} = 5$. Note that since this value is positive, it would correspond to an elasticity of E<1 indicating that the situation is inelastic. The "5" would indicate that a change in price p of $1 would yield roughly a change in revenue R of $5.

On the other side of the coin, suppose $\dfrac{\Delta R}{\Delta p} = -5$. This would correspond to an elasticity E>1 indicating that the situation is elastic. Thus, a price increase of $1 would yield a revenue increase of -$5, which is a revenue decrease.

Suppose the elasticity is 1. As you know, 1 is the magic number. Anything less than this and the situation is inelastic and revenue and price move in the same direction. Anything more than this and the situation becomes elastic and revenue and price move in opposite directions. Suppose E=1 and you lower the price. What happens to revenue? Since you have lowered the price, the situation will become inelastic so this drop in price will result in a drop in revenues. Suppose you raise the price. This will make the situation become elastic, and since you raised the price, revenue will again drop. Thus, you have a lose-lose situation. Two conclusions can be drawn from this:

1. An elasticity of E=1 corresponds to the maximum revenue, and
2. When the elasticity is hovering around 1, no dramatic price changes should be made without some external reason.

Information needed to work exercise 5.3

1. Formulas for elasticity of demand:

$$\text{Inelastic: } E = -\frac{\frac{\Delta q}{(q_1 + q_2)}}{\frac{\Delta p}{(p_1 + p_2)}} < 1 \text{ if and only if } \frac{\Delta R}{\Delta p} > 0$$

$$\text{Elastic: } E = -\frac{\frac{\Delta q}{(q_1 + q_2)}}{\frac{\Delta p}{(p_1 + p_2)}} > 1 \text{ if and only if } \frac{\Delta R}{\Delta p} < 0$$

2. If E=1 (or close), revenue will be at or near a maximum.

3. Since we can cancel the 2's in the denominator, the midpoints formula for elasticity

$$-\frac{\frac{\Delta q}{\left(\frac{\text{old } q + \text{new } q}{2}\right)}}{\frac{\Delta p}{\left(\frac{\text{old } p + \text{new } p}{2}\right)}} \text{ reduces to } -\frac{\frac{\Delta q}{\text{old } q + \text{new } q}}{\frac{\Delta p}{\text{old } p + \text{new } p}}. \text{ Since } \Delta q = q_2 - q_1 = q_{\text{new}} - q_{\text{old}},$$

calculator use of this formula should be done using the elasticity equation in the form:

$$E = -\frac{\left(\frac{(q_2-q_1)}{(q_2+q_1)}\right)}{\left(\frac{(p_2-p_1)}{(p_2+p_1)}\right)}$$

Exercise 5.3

1. Since fish spoils so quickly, it is important to a storekeeper that the price be slightly inelastic. Even though more revenue could be made by raising the price, unsold fish quickly spoils so there would be an overall loss in profit if the price were to become elastic. A storekeeper observes the following price-quantity relationship:

	Monday	Thursday
Price	$3.89	$3.39
Quantity	100 lbs.	110 lbs.

   a. What is the elasticity of demand?

   b. What should the storekeepers strategy be? and why?

2. Since sheet rock has few suppliers, is expensive to import due to weight considerations and is in constant if fluctuating demand, producers need to keep a close watch on the elasticity of demand. A recent price change produced the following results:

	NOVEMBER 90	MARCH 91
Price	$10.50	$12.00
Quantity	1.75 units	1.31 units

   a. What is the elasticity of demand?

   b. What should the strategy be? and why?

3. An acquaintance of mine ran an ice cream store for several years, never quite making enough profit to make it self-supporting. Since his franchise contract was up for renewal, and having just learned about elasticity in this class, he decided to check on the elasticity of ice cream sales. Paul's Ice Cream Parlor ran an advertised special. For one week only, all dips are 10¢ off. Hence, the 65¢ cone is 55¢ and the double dipper $1.20 is $1.00. At the end of the week Paul discovers that his sales are the same as they were the previous weeks.

   a. What is the elasticity?
   b. What is Paul's conclusion?

4. Northfone, the phone company for a remote exchange, charges 10¢ per call on their pay telephones. At this rate they find that about 7% of the local population use the pay phone each month. After studying data from a similar area with a 25¢ charge, they discover that about 6.5% of that population uses the pay phones each month. What should Northfones strategy be and why?

5. A company has gathered the following (short term) data on sales of a garden implement:

price	23.99	19.99	25.99	18.99	29.99
quantity	550	680	470	700	80

   a. If we consider only revenue, what pricing strategy would seem to be the best?
   b. The marginal cost of the item is $15. Now what do you think?

6. From observing sales data a merchant has drawn the following conclusions about a product sold:

SEASON	WINTER	SPRING	SUMMER	FALL
START SEASON #	1	4	7	10
price	$42	$42	$42	$42
elasticity	5	2	.75	1.5

a. What kind of a curve might best fit this data (season vs. elasticity)? Explain your choice.

b. Give the elasticity for each of the months May (5), June (6) and July (7). How does your predicted elasticity for July compare to the observed elasticity of 0.7? Did you choose a good function? Why or why not.

7. Given the following sales changes for a period of months at the different pricing levels, graph the elasticity, say what the best price is for each month, and indicate on the elasticity graph how you arrived at these conclusions.

Month	Oct.	Nov.	Dec.	Jan.	Feb.	Mar.
$39.50	50	275	1550	23	51	55
$29.50	55	350	2200	35	60	72
elast.						

A Review: Applying your knowledge

A company observes the following sales-price relations (in each column, prices are fixed at the start of the month, sales (quantity) are recorded at the end of the month).

A	B	C	D	E
1 Month	1	2	3	4
2 price	3.82	3.9	4	4
3 Sales (q)	32003	31900	32300	32400

	F	G	H	I	J	K	L
1	5	6	7	8	9	10	11
2	4	4.2	3.9	4	4	3.9	4.1
3	32600	29506	31050	31500	31400	32700	32900

Find and graph the following:

a. Calculate and graph the revenue function.

Given that the marginal cost is $3.85 per item (disregard fixed cost), calculate and graph

b. The profit function P.

c. An estimate for the marginal profit, $\frac{dP}{dq}$, and

d. The rate of change of the marginal profit.

e. Using the information you have about the profit, marginal profit and rate of change of marginal profit (the discrete form of the second derivative), what conclusions can you draw about the proper price of the item? GIVE AT LEAST ONE CAREFULLY THOUGHT OUT REAL WORLD EXAMPLE TO SUPPORT YOUR ANSWER.

f. Assume a fixed cost "k" and use the break-even point to find the largest fixed cost k you could have and still turn a profit during each time period. (Revenue≥(marginal cost)*q+k when k≤Revenue-(marginal cost)*q.)

g. Scanning across the results, what can you say about the fixed cost?

h. What affect does the second derivative of the profit have on the profit function?

i. When is a 0 second derivative good? When is a 0 second derivative bad? Why would I say the points of the function are like the "ghost of Christmas past", the first derivative is like the "ghost of Christmas present", and the second derivative like the "ghost of Christmas future"?

Section 5.4 Point Elasticity Of Demand

Let's return now to the original definition of the derivative. You will recall that if we define price p, in terms of quantity q, we get a function p(q) where price depends on quantity. Using such a function, our derivative quotient is: $\frac{p(q+h)-p(q)}{h}$. Another way to express this derivative quotient is to realize that p(q+h)-p(q) is the change in p, which we have been calling Δp, and "h" is the change in q or Δq; thus the quotient $\frac{p(q+h)-p(q)}{h}$ can be expressed as the quotient $\frac{\Delta p}{\Delta q}$. Now recall that we actually calculated the derivative by taking the $\lim_{h \to 0} \frac{p(q+h)-p(q)}{h} = \frac{dp}{dq}$. Since h corresponds to Δq, letting h approach 0 is the same as letting Δq approach 0. Thus the equation $\lim_{h \to 0} \frac{p(q+h)-p(q)}{h}$ can be rewritten as $\lim_{\Delta q \to 0} \frac{\Delta p}{\Delta q} = \frac{dp}{dq}$. This notation can be very handy. For example, we must use not only the ratio $\frac{\Delta p}{\Delta q}$ but also the ratio $\frac{\Delta q}{\Delta p}$. Calculating the limit as Δp approaches 0 we get $\lim_{\Delta p \to 0} \frac{\Delta q}{\Delta p} = \frac{dq}{dp}$. Note that this notation reveals which is considered the dependent and which is considered the independent variable. In the equation, $\frac{dp}{dq}$ we see that p is a function of q, ie., p is the dependent and q the independent variable. In the equation $\frac{dq}{dp}$, p is the independent and q the dependent variable. For a couple of simple examples to see how this works in practice, consider the following:

Example 1: Linear demand functions

1. p=1-q
2. p=1+3q

In equation 1, $\frac{dp}{dq}=-1$, and in equation 2, $\frac{dp}{dq}=3$. Note that both equations are solved for p in terms of q, ie., p is the dependent and q is the independent variable. Now let us calculate $\frac{dq}{dp}$. To do this, we will first solve each equation for q in terms of p.

1. q=1-p
2. $q = \frac{p-1}{3} = \frac{p}{3} - \frac{1}{3}$

In equation 1, $\frac{dq}{dp} = -1$ and in equation 2, $\frac{dq}{dp} = \frac{1}{3}$.

Do you see any relationship between the two sets of answers? There is a very important one which can easily be seen by looking at the fractions we started with, $\frac{\Delta p}{\Delta q}$ and $\frac{\Delta q}{\Delta p}$. Observe that $\frac{\Delta q}{\Delta p} = \frac{1}{\frac{\Delta p}{\Delta q}}$. Thus it will be the case that:

Relation between $\frac{dq}{dp}$ and $\frac{dp}{dq}$: $\quad \frac{dq}{dp} = \frac{1}{\frac{dp}{dq}}$ and $\frac{dp}{dq} = \frac{1}{\frac{dq}{dp}}$

This theorem can be of great importance in dealing with point elasticity. It is sometimes the case that we can actually come up with an equation for the demand function for some product. If this happens we can find the **point elasticity of demand**. The difference between the point elasticity of demand and the elasticity we discussed in the last section is that the elasticity requires values at two distinct points whereas the point elasticity of demand is calculated at a single point. As you might guess, calculating the point elasticity of demand requires the use of calculus. Fortunately, using our newly developed notation considerably simplifies our task.

Look again at the elasticity function, $E = -\dfrac{\dfrac{\Delta q}{(q_1 + q_2)}}{\dfrac{\Delta p}{(p_1 + p_2)}}$. Recall that when we developed this equation, both the $q_1 + q_2$ and $p_1 + p_2$ were divided by 2 (we canceled the 2's make rewriting and calculations easier). Placing the 2's back into the equation and changing q_1 and p_1 to just "q" and "p" we get

$$-\dfrac{\dfrac{\Delta q}{\left[\dfrac{(q + q_2)}{2}\right]}}{\dfrac{\Delta p}{\left[\dfrac{(p + p_2)}{2}\right]}}$$

Inverting the denominator and multiplying yields:

$$-\frac{\Delta q}{\left(\frac{q+q_2}{2}\right)} * \frac{\left(\frac{p+p_2}{2}\right)}{\Delta p}$$

We may now rearrange the terms as follows:

$$-\frac{\Delta q}{\Delta p} * \frac{\left(\frac{p+p_2}{2}\right)}{\left(\frac{q+q_2}{2}\right)}$$

We now wish to take the limit as Δq approaches 0. Consider the first fraction. We have already observed that $\lim_{\Delta q \to 0} \frac{\Delta p}{\Delta q} = \frac{dp}{dq}$ and invoking our theorem, we see that $\lim_{\Delta q \to 0} \frac{\Delta q}{\Delta p} = \frac{dq}{dp}$. In the case of the second fraction, you have to realize what happens to the quantity $(q+q_2)/2$ as Δq approaches 0. As Δq approaches 0, the midpoint determined by $(q+q_2)/2$ slides back towards p, and when $\Delta q = 0$ that point is q. Similarly, the midpoint $(p+p_2)/2$ slides back towards p and when Δq is 0, Δp is also 0 and the midpoint $(p+p_2)/2$ becomes the point p. Thus,

$$\lim_{\Delta q \to 0} -\frac{\Delta q}{\Delta p} * \frac{\left(\frac{p+p_2}{2}\right)}{\left(\frac{q+q_2}{2}\right)} = -\frac{dq}{dp} * \frac{p}{q}$$

This equation, $-\frac{dq}{dp} * \frac{p}{q}$, is called the "point elasticity of demand" and is signified by the greek letter eta (η):

Point elasticity of demand:

$$\eta = -\frac{dq}{dp} * \frac{p}{q}$$

Just as we observed using discrete data, we have the following results using a differentiable function:

POINT ELASTICITY OF DEMAND:

$$\eta = -\frac{dq}{dp} * \frac{p}{q}$$

If $\eta > 1$ the demand is elastic.
If $\eta < 1$ the demand is inelastic.

Example 2: Point elasticity example

Given a demand function p=12-2q, what is the elasticity when p=$6?

solution: dp/dq=-2, so (by our theorem) dq/dp=-1/2. When p=6, q=3. Hence, $-\frac{dq}{dp} * \frac{p}{q} = \frac{1}{2} * \frac{6}{3} = 1$.

USING CALCULUS: THE SPEEDY APPROACH:

The use of calculus permits a quick and artificial way of arriving at the formula $\eta = -\frac{dq}{dp} * \frac{p}{q}$. While you will enjoy the speed with which we develop this formula, note that as we develop elasticity this way it does obscure the fact that the midpoints formula is crucial when working with discrete data.

For the speedy approach, start with the fact that revenue=price*quantity:

$$R = pq$$

Consider the equation to be a function of "p" and differentiate using the product rule:

$$\frac{d(R=pq)}{dp} \text{ yields } \frac{dR}{dp} = \frac{dp}{dp} * q + p * \frac{dq}{dp}$$

and (dp)/(dp) (the derivative of p with respect to p) =1.

Now recall that when demand price is elastic, revenue and price move in opposite directions and $\frac{dR}{dp} < 0$. Thus, $q + p * \frac{dq}{dp} < 0$ or $p * \frac{dq}{dp} < -q$. Dividing both sides by -q (the inequality has to change directions since -q is negative), we get:

$$-\frac{p}{q} * \frac{dq}{dp} > 1.$$

Similarly, if the demand price is inelastic, $-\frac{p}{q} * \frac{dq}{dp} < 1$. Hence we again come up with the same equations for point elasticity of demand:

$$\eta = -\frac{dq}{dp} * \frac{p}{q}$$

Some of the things we have been considering can be somewhat simplified by invoking some calculus considerations. Recall that $\frac{dR}{dp} = q + p * \frac{dq}{dp}$ and the discrete approximation to this is $\frac{\Delta R}{\Delta p} = q + p * \frac{\Delta q}{\Delta p}$. Multiplying through by Δp we obtain $\Delta R = q * \Delta p + p * \Delta q$ and factoring out $p*q$ we get: $\Delta R = pq\left(\frac{\Delta p}{p} + \frac{\Delta q}{q}\right)$. Since R=pq, we can divide the left side by R and the right side by pq yielding $\frac{\Delta R}{R} = \frac{\Delta p}{p} + \frac{\Delta q}{q}$. Finally, since $\frac{\Delta p}{p}$ is the ratio of the change in price, if this percentage change is 1, (ie., $\frac{\Delta p}{p} = 1$) we get $\frac{\Delta R}{R} = 1 + \frac{\Delta q}{q}$ and looking at $\eta = -\frac{dq}{dp} * \frac{p}{q}$ and substituting 1 for $\frac{\Delta p}{p}$, we get:

$$\frac{\Delta R}{R} = 1 - \eta$$

This can be a handy formula to use, if you will be careful. First, note that we let the **percentage** change be 1. This means that this equation is a "per cent" equation, ie., work the entire problem in per cents. To see how easy it is to use this equation under some circumstances, look at the following example.

Example 3: Using η to find $\frac{\Delta R}{R}$

The elasticity for a certain product at the current price is known to be .8. If the price is raised by 5%, what will the change in revenue be?

solution: Since the price is raised by 5% and $\eta = .8$, substitute .8 in the equation 1-η

and multiply by 5: (1-η)*5=1. The revenue will increase by 1%.

Exercise 5.4:

1. If short term elasticity for a product is known to be 2 at a certain price and the price is dropped 10%, what is the per cent change in the revenue?

2. If the demand function for peanuts is $p = \dfrac{30q + 7000}{q + 100}$, find:
   a. The market equilibrium point.
   b. The point elasticity of demand when p=$50.
   c. Given a cost function of C(q)=35q+300, find the production that maximizes profit.

3. The demand function for a certain commodity is $p=60-q+.004q^2$ and the supply function is $p=10+.3q-.002q^2$, $0 \leq q \leq 75$. Find:

   a. The market equilibrium point.
   b. The point elasticity of demand when p=$22.
   c. Given a cost function of C(q)=8q+500, find the production that maximizes profit.

Section 5.5 Further Graph Analysis: The interaction between a function and its first and second derivatives (Optional)

A curve is said to be **concave up** if it opens upward; it is **concave down** if it opens downward. To relate the concept of concavity to tangent lines, let us look at the following diagrams.

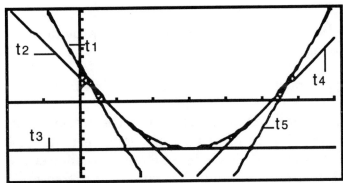

Relationship: Concave up and the tangent lines

Attaching (guessing) some numerical values to the slope of the tangent lines we might get:

slope t_1 = -7, slope t_2 = -5, slope t_3 = 0, slope t_4 = 3, slope t_5 = 6

Note: **The slope of the tangent line increases as the point of tangency moves to the right.**

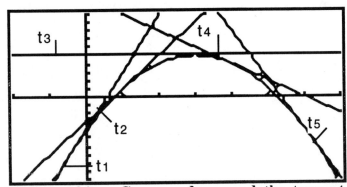

Relationship: Concave down and the tangent lines

Attaching some numerical values to the slope of the tangent lines we might get:

slope t_1 = 8, slope t_2 = 1, slope t_3 = 0, slope t_4 = -1, slope t_5 = -7

Note: **The slope of the tangent line decreases as the point of tangency moves to the right.**

We might state or make the following observations:

1. The graph of the function is said to be concave up on an interval provided f' (slope of tangent lines) is increasing.

2. The graph of the function is said to be concave down on an interval provided f' (slope of the tangent lines) is decreasing.

Assuming that our observations are correct, how might we find these intervals of x on which f' is increasing or decreasing?

Recall that the first derivative of f (denoted $f'(x)$) indicates where the graph of f is increasing and where it is decreasing; therefore, the derivative of f' (the derivative of the derivative - denoted $f''(x)$) should indicate where f' is increasing and where it is decreasing. $f''(x)$ is called the **second derivative** . Using the concept of the second derivative, we can now make the following observations:

Observation 1:

A positive second derivative corresponds to those places where the original function is concave up.

Observation 2:

A negative second derivative corresponds to those places where the original function is concave down.

Example 1: Quadratic functions

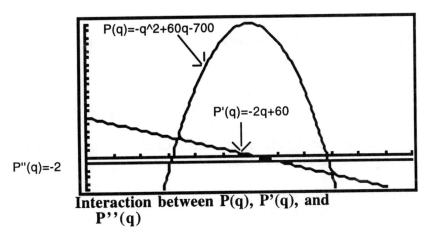

Interaction between P(q), P'(q), and P''(q)

Consider the case where the original function is a quadratic. Let $P(q) = -q^2 + 60q - 700$. Then $P'(q) = -2q + 60$ and $P''(q) = -2$. Notice that the second derivative in this case is always negative and that the concavity of this quadratic function is always down.

Example 2: Polynomial - degree 3

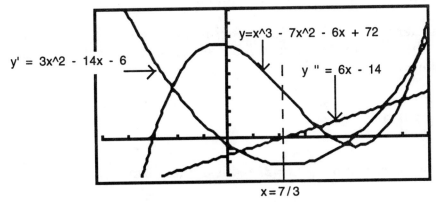

Interaction between y, y', y''

The second derivative ($y'' = 6x-14$) has a positive value for x's greater than $\frac{14}{6} = \frac{7}{3}$. Notice that this is also the interval where y' is increasing and where y is concave up. The second derivative has a negative value for x's less than $\frac{14}{6} = \frac{7}{3}$. Notice this is where y' is decreasing and where y is concave down. Looking at this example, we see that $y'' = 0$ when $x = 7/3$; therefore we might make another observation:

Observation 3:

A zero second derivative often corresponds to those places where the original function reverses concavity.

Second derivative theorem:

If a function g is at least twice differentiable (do not talk about the second derivative if it does not have one), and its second derivative is continuous,

1. $g'' > 0$ corresponds to the region of g which is concave up,
2. $g'' < 0$ corresponds to the region of g which is concave down,
3. if g'' changes signs as it passes through $g'' = 0$, $g'' = 0$ corresponds to the point where the concavity changes and is called an **inflection point**.

Example 3: Linear second derivatives

Given a function with second derivative $2x-4$ and the fact that the original function passes through the point (0,0), sketch the original function.

solution: Using the second derivative as our starting point, we calculate a few values for it. How do you choose the x-values to be used for evaluation purposes? You are interested in three general areas:

1. Those values where $g'' < 0$
2. Those values where $g'' > 0$, and
3. Those values where $g'' = 0$.

In this first example our work is easy since $g''(x) = 2x - 4$ is a linear function.

1. $2x-4 < 0$ when $x < 2$.
2. $2x-4 > 0$ when $x > 2$, and
3. $2x-4 = 0$ when $x = 2$.

Thus, we need to look at a few values less than 2 and a few greater than 2.

x	-5	-2	0	2	5	10
2x-4	-14	-8	-4	0	6	8

In general, looking at the values of the second derivative, we see that

1. from -∞>x>2, g(x) is concave down,
2. from 2<x<∞, g(x) is concave up, and
3. at x=2 g(x) changes concavity.

We use the second derivative and the fact that our function passes through (0,0) to give us a rough idea of what the original function looks like.

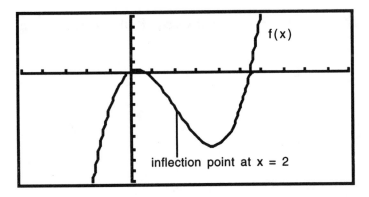

inflection point at x = 2

As we shall see in a later chapter, when you are fortunate enough to have a nice second derivative like 2x-4 to start with, and a point that your original function passes through ((0,0) in this case), you will be able to <u>exactly</u> determine the original function. This fact will provide some valuable information at times.

Point of Diminishing Returns

In economics, a point of maximum rate of change is called the **point of diminishing returns.** This point of diminishing returns occurs at a point of inflection where the curve changes from concave up to concave down.

Example 4: Point of diminishing returns

Given the following data points where x represents thousands of dollars spent on advertising and S is sales in thousands of dollars, find a degree 3 polynomial regression curve of best fit. Sketch the graph and find the point of diminishing returns.

x	2	5	8	11
S(x)	227.2	330.25	479.2	633.55

Entering the data points in the STAT editor and finding the polynomial (degree 3) of best fit we get the following:

$S(x) = -.2500000000004x^3 + 6.300000000014x^2 - .00000000003x + 2040000000004$

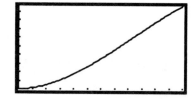 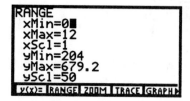

We can now use the INFLC feature of the GRAPH editor of our calculator (MATH, MORE) to find the point of inflection.

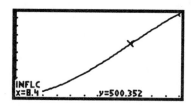

This inflection point is the point of diminishing returns

Information needed to work exercise 5.5

1. **Second derivative theorem** : If a function g is at least twice differentiable and its second derivative is continuous,

   1. g">0 corresponds to the region of g which is concave up,
   2. g"<0 corresponds to the region of g which is concave down,
   3. if g" changes sign as it passes through g"=0, g"=0 corresponds to the point where the concavity changes and is called an **inflection point**.

2. A point of diminishing returns occurs at a point of inflection where the curve changes from concave up to concave down.

Exercise 5.5

Find all points of inflection

1. $y = 3x^4 - 4x^3 + 3$

2. $y = x^3/3 - 4x + 2$

3. Given the following data points where x represents thousands of dollars spent on advertising and S is sales in thousands of dollars for automobile tires, find a degree 3 polynomial regression curve of best fit. Graph this regression curve and find the point of diminishing returns.

x	10	25	50	75
S(x)	380	1662.5	5100	8537.5

Section 5.6 Using Calculus Concepts On Discrete Data (Optional)

It would certainly be a wonderful thing if data came complete with an accompanying continuous function; if your accountant would report profit and loss as differentiable functions, but alas, that is not the case. Functional notation is simply not a part of the everyday world where one makes observations daily, weekly, monthly, yearly, etc. The correct descriptive term for such observations is **discrete**. Each observation made is separate from the others.

Thus, instead of a continuous function, our observations yield a finite collection of distinct and separate points. Most of the time it is not possible to say exactly what the original function, the marginal function (the first derivative), or the rate of change of the marginal function (the second derivative) is. This will not prevent us from analyzing the situation as we shall see in the next example.

Since such observations do not yield differentiable functions, one cannot use the differential calculus that we have learned so far on them. However, we can apply the **concepts** we have learned.

Example 1: Estimating derivatives with difference quotients

Consider the following set of sales totals (revenues) and the corresponding years:

1985	1986	1987	1988	1989	1990	1991	1992	1993
1	1.5	2.5	4	6	8.5	11.5	15	19

We can see that sales have been increasing at a healthy pace. Taking only this into account, everything looks great.

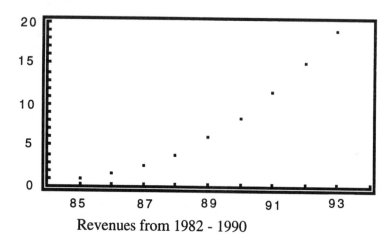

Revenues from 1982 - 1990

However, it is often the case that the first derivative can reveal a trend which is not so obvious. Since we do not have an equation for whatever the function might be, we will need to look at the difference quotient, $\frac{f(x+h) - f(x)}{h}$, where g is the revenue function and h is the increment by year. This ratio $\frac{\text{change in revenue}}{\text{change in years}}$ will approximate $\frac{dR}{dt}$. There is a notation for this "change" that has been used for several centuries. The capital greek letter Δ. Using this change notation, our fraction

$$\frac{\text{change in revenue}}{\text{change in years}}$$

can be expressed as

$$\frac{\text{change in revenue}}{\text{change in years}} = \frac{\Delta R}{\Delta t}.$$

We will frequently use this notation. In estimating $\frac{dR}{dt}$ with $\frac{\Delta R}{\Delta t}$ we need to make each value as small as possible. This means that at each point (year) we will let the change Δt be as small as possible, but not zero. Hence $\Delta t = 1$ is as small as it can get. For example, looking at the data and letting $\Delta t = 1$, $\frac{F(1985+1) - f(1985)}{1} = \frac{1.5 - 1}{1} = .5$.

Since $\Delta t = 1$, our calculation for $\frac{\Delta R}{\Delta t}$ will be very simple. We need only find the value of ΔR which is the difference between any two successive revenue entries.

Filling in the remaining table we get:

1985	1986	1987	1988	1989	1990	1991	1992	1993
1	1.5	2.5	4	6	8.5	11.5	15	19
ΔR	.5	1	1.5	2	2.5	3	3.5	4

Note that the first entry is blank, since each entry is determined by subtracting the entry from the previous entry and the first entry (1985) has no prior entry.
Plotting both sets of data on the same graph, we see that the slope of the data is a straight line with positive slope.

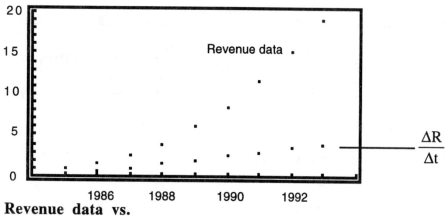

**Revenue data vs.
(change in Revenue)/(change in time) data**

This indicates that not only are sales increasing, the **rate** at which they are increasing is itself increasing. Looking at the difference quotient of the difference quotient (an estimate of the second derivative of the function), we see that the value is constantly equal to .5 indicating both that the first difference quotient is increasing and that the function is concave up.

Estimate of second derivative:

1985	1986	1987	1988	1989	1990	1991	1992	1993
1	1.5	2.5	4	6	8.5	11.5	15	19
ΔR	.5	1	1.5	2	2.5	3	3.5	4
change in ΔR		.5	.5	.5	.5	.5	.5	.5

Happy days are forever! (?)

Example 2: Using calculus concepts to look at discrete data trends

Suppose now we had considerably more data and the situation were not as simple as the last problem. Taking a small sample of the available data we observe the following:

units	2	36	101	151	201
profit	-$69.39	$9.72	$75.15	$48.82	-$44.18

Given only the numbers in our table, we see that everything looked fine until we hit the 201st item, at which point we were suddenly losing money again. Was there any **PRIOR** warning that we would start losing money at about the 201st item? The answer in this case is a solid yes. Look at the following observed data sets. P stands for recorded profit. MP is the calculated marginal profit, P(q+1)-P(q), and MMP stands for the rate of change of marginal profit, MP(q+1)-MP(q), the discrete form of the second derivative of the profit.

To get enough data to work with we will need extra data in sets of three values. The reason for this is that to find the rate of change of the marginal profit (2nd derivative) we will need two values of the marginal profit (1st derivative), and to find two values of the marginal profit we will need three values of the profit function. Thus, to discuss what is happening to the second derivative around quantity=2 we will need a minimum of three values:

quantity	1	2	3
Profit=P	-72.18	-69.39	-66.62
Marg. Profit=MP		2.79	2.77
change in MP= MMP			-0.027

Collecting the necessary data and doing our calculations we get:

	1	2	3	35	36	37
P	-72.18	-69.39	-66.62	7.83	9.72	11.58
MP		2.79	2.77		1.89	1.86
MMP			-0.027			-0.027

	100	101	102	150	151	152
P	75	75.15	75.28	50	48.82	47.61
MP		0.15	0.13		-1.18	-1.21
MMP			-0.027			-0.027

	200	201	202			
P	-41.67	-44.18	-46.72			
MP		-2.51	-2.54			
MMP			-0.027			

It only takes a moment to notice in the table that the marginal profit is a decreasing function throughout our range of interest. What does this mean? This decreasing marginal profit is a reflection of the fact that the rate of change of the marginal profit (the second derivative) is constantly -0.027. Thus, with this constant negative slope of the marginal profit, the marginal profit will eventually become negative. The instant this happens, profit will have reached its peak and be on a downward slide.

If you were watching this data as it was recorded, a reasonable conclusion would be that the first time the marginal profit became negative, hard times would not be far off because of the apparent intransigence of the rate of change of the marginal profit.

Example 3: Looking at business index functions

The following is typical of the reports issued by economists as they analyze the economy. In constructing such reports, they usually "seasonally adjust" the raw numbers to reflect the effect of the time of the year and often use "smoothing" techniques to remove some of the variation which can cause variation in the data they do not wish to study.

The report:

Economic activity remained sluggish in December, according to an index of economic indicators. The Index dipped slightly in December. The decrease stemmed from a drop in total employment and retail sales during the month. Total employment dropped to 590,000 from 590,700 in November. The unemployment rate remained unchanged at 6.5 percent. Retail sales dropped to $728.1 million from $739 million. The value of residential building permits issued, a reflection of plans for future construction, rose to $34 million in December from $28.7 million in October and $32.3 million in November. The total value of permits issued in December was up 39.3 percent when compared with the same period a year ago.

Business Index - monthly percentage change					
J	F	M	A	M	J
-.4%	-.15%	-.06%	.15%	.27%	.2%

J	A	S	O	N	D
.07%	.14%	.07%	.07%	0%	-.07%

Assignment:

1. Generate the index function from its derivative (the change function), then graph the change function vs. the index function.
2. Generate the second differences of the index function (the rate of change of the change function) graph the second differences vs. the index function.
3. Given the data as stated in the article, the values of the second differences and the index function, what is your opinion about the index for the next two periods? Do you think it will rise, go down, or stay the same.

Solution:

To reconstruct the index function we will start with the per cent change, arbitrarily assign the index function the value "1" at the start and apply the change function to it. Thus, if the index function started at the value one and increased in January by -.4%, it decreased by .004, ie., 1+(-.004)(1)=0.996. In February the increase was -.15% or -.0015, so the value of the index in February was .996+(-.0015)(.996)=.9945.

For the remainder of the calculations are as follows:

Note: We could enter the equation I=PI + R* PI in the SOLVER. Remember to change the % to a decimal before substituting for R. The PI stands for previous index.

The March entry can be found as follows:

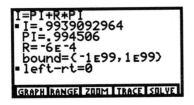

Now, substitute for **PI** the value of **I** for March and enter .0015 for R to get the April index.

Business Index - monthly percentage change								
Month	t	J	F	M	A	M	J
change %	ΔI/Δt %	-.4%	-.15%	-.06%	.15%	.27%	.2%
I=index	1.0	.996	.9945	.9939	.9954	.9981	1

J	A	S	O	N	D
.07%	.14%	.07%	.07%	0%	-.07%
1.0008	1.0022	1.0029	1.0036	1.0036	1.0029

Graphing these two functions:

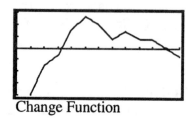

Change Function · Window Setting

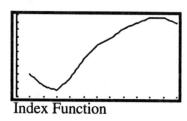 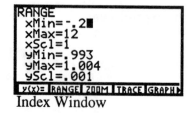

Index Function · Index Window

Storing these pictures in the DRAW menu of the STAT editor and recalling the Change function on top of the Index, we get the following:

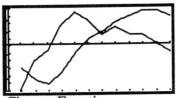
Change Function vs. Index Function

Now to consider the second differences or rate of change of the change function. When viewing the second differences of a function it is usually sufficient to see the sign. Hence, we will use the per cent values given in the change function to find the rate of change of the change function. Since the value for January is -.4 and the value for February is -.15, the second derivative would be (-.15-(-.4))=+.25%. Continuing in this fashion we get:

	Business Index - monthly percentage change					
	J	F	M	A	M	J
$\Delta I/\Delta t$ %	-.4%	-.15%	-.06%	.15%	.27%	.2%
	1.0	.996	.9945	.9939	.9954	.9981	1
change in	$\Delta I/\Delta t$ %	.25	.09	.21	.12	-.07

J	A	S	O	N	D
.07%	.14%	.07%	.07%	0%	-.07%
1.0008	1.0022	1.0029	1.0036	1.0036	1.0029
-.13	.07	-.07	0	-.07	-.07

Graphing the second derivative we get the following:

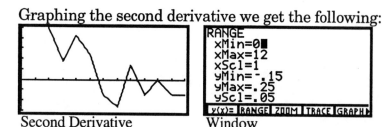

Second Derivative   Window

Graphing the second differences against the index function we see the concavity of the index function is dramatically connected to the sign of the second differences, the discrete form of the same result for second derivatives.

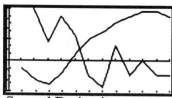

Second Derivative vs.
Index Function

Finally, what, if any, conclusions can we or should we draw from this information? You may observe that the value of the second differences for the last four periods is very small (-.0007) and seems to be fluctuating a little. Gleaning the significant information from the news article:

a. Total employment dropped to 590,000 from 590,700 in November.
b. The unemployment rate remained unchanged at 6.5 percent.
c. Retail sales dropped to $728.1 million from $739 million.
d. The value of residential building permits issued, a reflection of plans for future construction, rose to $34 million in December from $28.7 million in October and $32.3 million in November. The total value of permits issued in December was up 39.3 percent when compared with the same period a year ago.

Knowing that building permits are one of the leading economic indicators (recall our first large assignment), noting that an employment drop of 700 is not significant, and that retail sales is a reflection more of what has happened than what is going to happen, we could logically conclude that within the next two periods, the index should rise, perhaps significantly.

© 1995 Saunders College Publishing

Information needed to work exercise 5.6

1. When using discrete data, the derivative of a function f(x) with respect to x corresponds to the quotient . The "change in x" cannot be smaller than the situation allows, eg., if x is in items, then 1 is as small as the denominator can get.
2. The second derivative (second differences) of a function is the derivative (differences) of the first derivative (first difference). When analyzing data it is often helpful to look at the signs of the second derivative (second differences) to determine concavity.

+ yields concave up, - yields concave down

Exercise 5.6

1. Economic activity fell in January for the third consecutive month, according to an index of economic indicators. The Business Index dropped 0.02 percent in January because of a drop in total employment, a decline in the value of residential building permits issued and lower earnings for manufacturing workers. Employment dropped by 1,500, from 597,100 in December to 595,600 in January. The unemployment rate dropped from 6.1 percent in December to 5.0 percent in January despite the decline in total employment. The value of residential building permits issued, which is a reflection of construction activity, dropped from $52 million to $51.4 million. Retail sales rose from $721 million in December to $723.5 in January.

Business index - percentage change						
F	M	A	M	J	J
-.33%	-.28%	-.33%	-.28%	-.13%	1.5%

A	S	O	N	D	J
.4%	.4%	0%	-.13%	-.13%	-.02%

Assignment:

a. Generate the index function from its the change function, then graph the change function vs. the index function.
b. Generate the approximation to the second derivative of the index function (the rate of change of the change function). Graph the approximation of the second derivative vs. the index function.

c. Given the data as stated in the article and your graphs, write a paragraph giving your opinion about the index for the next two periods and the general trend of the things mentioned in the article. Your statements should be directly connected to your graphs.

Comments on your graphs should include the connections between the index and the marginal index at appropriate points (graph a.), and comments about the connection between the change in the marginal index and the index (graph b.).

2. A furniture company observes that their sales are extremely sensitive to the economy, with sales usually following almost exactly the trend of an index of leading economic indicators by 1 month. If the index drops by more than 0.2% and stays down that much for a month, their sales will respond by dropping nearly 100 times that amount. The index has recorded the following per cent changes for the last six weeks. The company's current revenues are about 2.5 million per week.

% change in index					
-.1%	-.1%	-.04%	.01%	0%	.01%

a. Superimpose the graph of the index function vs. the change in the change in the index function (approximation to 2nd derivative of index).
b. Using only the graph, predict the index for the next month. Recalling earlier problems, what additional information would be crucial in making this prediction?
c. Superimpose the graph of the revenue vs. index on the same scale by adjusting the revenue function.
d. Write a paragraph explaining what the company can expect. Be sure to connect your statements to your graphs. You should write some of what you have to say on your graphs in the appropriate places.
e. Write a paragraph explaining what the company should do. Consider different cost cutting activities such as vacations, reduced work week (4 days) reduced work hours, lay-offs, and potential revenue increasing activities such as sales, etc.

3. Economic activity fell in November for the second consecutive month, according to an index of economic indicators. The Business Index dropped 0.1 percent in November because of the drop in total employment, a decline in the value of residential building permits issued and lower earnings for manufacturing workers. Employment dropped by 1,000, from 493,000 in October to 492,000 in November, with the decline mostly in manufacturing jobs. The unemployment

rate increased from 5.0 percent in October to 5.1 percent in November. The value of residential building permits issued, which is a reflection of construction activity, dropped from $28 million to $25 million. Retail sales rose from $621 million in October to $709 in November.

Business index - percentage change					
D	J	F	M	A	M
-.2%	-.18%	-.4%	-.1%	-.02%	.2%

J	J	A	S	O	N
.3%	.4%	.1%	0.0%	-.13%	-.02%

Assignment:

a. Generate the index function from its change function, then graph the change function vs. the index function.
b. Generate the approximation to the second derivative of the index function (the rate of change of the change function) graph these values vs. the index function.
c. Given the data as stated in the article and your graphs, write a paragraph giving your opinion about the index for the next two periods and the general trend of the things mentioned in the article. Your statements should be directly connected to your graphs, writing necessary comments on your graphs as you work.

Graphs with no comments written on them and/or paragraphs with no specific graph references and no specific article references, will not be acceptable. Comments on your graphs should include the connections between the index and the marginal index at appropriate points (graph a.), and comments about the connection between the change in the marginal index and the index (graph b.).

4. Consider again our construction industry problem. Analyze the problem again in light of your knowledge of the second derivative:

The construction industry is often watched closely to see if construction is increasing or decreasing (seasonally adjusted). This is used as a leading economic indicator to see if the economy is improving or not. Given the following data, make a prediction of the spending level for the next two months. Numbers are rounded off to the nearest billion dollars. Using these numbers, if the assumption of the relation is correct, what do you predict the construction industry (and hence the economy) will do in November and December?

	1990		1991									
Mon.	N	D	J	F	M	A	M	J	J	A	S	O
Amt	435	421	408	410	401	408	399	398	400	402	408	411

Your task is to find a curve of best fit and, using it, predict what will happen the next two months.

5. a. Given the following change function, recreate the index function. Use 1 as your starting value for the index function.
   b. Now calculate marginal change function from the change function.
   c. Now graph the marginal change function against the index function and hand draw what you get.
   d. What conclusions can you make? Write them out carefully.

Business index - percentage change

	Mar.	Apr.	May	June	July	Aug.
	.1%	.1%	-.04%	-.1%	.00%	-.2%
1						

6. a. Given the following change function, recreate the index function. Use 1 as your starting value for the index function.
   b. Now calculate marginal change function from the change function.
   c. Now graph the marginal change function against the index function and hand draw what you get.
   d. What conclusions can you make? Write them out carefully.

Business index - percentage change

	Mar.	Apr.	May	June	July	Aug.
	.1%	.1%	-.04%	-.1%	.00%	-.2%
1						

CHAPTER 6: INTEGRAL CALCULUS

Section 6.1 Antidifferentiation

Calculus involves two basic operations: differentiation and integration (or antidifferentiation). Up to this point we have looked at problems involving finding the derivative given the function. Many applications of calculus involve the inverse problem: **Given a marginal function (derivative), what was the original function or antiderivative that yielded this derivative?**

Definition of Antiderivative

A function F is an **antiderivative** of a function f if $F'(x) = f(x)$ for all x in the domain of f.

Example 1: Antiderivatives of linear functions

A simple example would be the following: If the marginal revenue is 3q, what is the revenue? An equivalent way of stating this question is to ask "what function has derivative 3q?"

The answer to this can be found after a moments reflection: $\frac{3q^2}{2}$. This can be checked by differentiating $\frac{3q^2}{2}$: $\frac{d\frac{3q^2}{2}}{dq} = 3q$. You need to observe one problem with this solution. Not only is $\frac{d\frac{3q^2}{2}}{dq} = 3q$, so is $\frac{d\frac{3q^2}{2} + constant}{dq}$. Since the derivative of any constant is 0, you may add any constant you wish, thus changing the original function. Under most circumstances we will have enough additional information to determine what this unknown constant is. Thus the process of antidifferentiation does not determinine a unique function, but a family of functions each differing by a constant.

This process of antidifferentiation is also called **integration** and it is denoted by this integral sign $\int$. Thus $\int f(x)\, dx$ denotes the **antiderivative of f with respect to x**. Therefore,

$$\int f(x)\, dx = F(x) + k,$$

where F is the antiderivative of f and k is an arbitrary constant.

This is called an **indefinite integral**.

For any linear function, aq+b, what is the antiderivative? We have already considered the antiderivative of aq when a=3. Note that the 3 did not enter into the solution. In the solution, you can actually disregard the 3 and look only at the exponent of q: q^1. The antiderivative was found by adding one to the exponent and dividing by the new exponent: $\int q^1\, dq = \frac{q^{1+1}}{1+1}$, for when you differentiate the result, $\frac{q^2}{2}$, you get q^1 as the answer. In finding the antiderivative of the "b" in aq+b you need only realize that the function bq has slope b. Thus, the general solution for the antiderivative of aq+b is $a\frac{q^2}{2} + bq + \text{constant}$.

One of our major tasks when solving such "reverse marginal" or **antidifferentiation** problems will be to find some reliable general techniques for finding the antiderivatives. That is, a few simple theorems would be of great help in deciding some of the more difficult questions. We have, in fact, just discovered one by observing that in finding the antiderivative of aq^1 only the exponent "1" actually was used to find the solution.

Example 2: Antiderivative of x^n

When trying to find antiderivatives it is crucial that you have a clear picture in your mind of all that you have learned about derivatives. Here is a quick and concise summary:

First, you have learned to take three elementary derivatives:

1. If $y = x^n$, then $\frac{dy}{dx} = nx^{n-1}$
2. If $y = \ln x$, then $\frac{dy}{dx} = \frac{1}{x}$, and
3. If $y = e^x$, then $\frac{dy}{dx} = e^x$

After refreshing your memory with these facts, consider the following pattern which is easily seen when dealing with **antiderivatives of x^n.**

1. Given the derivative is x^2, what is the function? Noting that the only way you can have a derivative with a power of 2 is to start with a derivative of power 3 ($y'=nx^{n-1}$), the derivative must have come from some function with an x^3 in it. Again observing that the derivative of x^3 is $3x^2$, not x^2, our original function must have been $\frac{x^3}{3}$, usually written $\int x^2\, dx = \frac{x^3}{3}$.

2. Given the derivative of a function is x^3, what was the original function? Making the same observations again, the original function must have had an exponent of 4, so it must have had an x^4 in it. Since the derivative of x^4 is $4x^3$, the original function must have been $\frac{x^4}{4}$. Thus, $\int x^3\, dx = \frac{x^4}{4}$.

3. Given the derivative of a function is x^n, what is the function? It is clear by now that the function must have been $\frac{x^{n+1}}{n+1}$. Hence, it appears that $\int x^n\, dx = \frac{x^{n+1}}{n+1}$

This almost gives us our first elementary antiderivative. I say almost because we have overlooked two problems.

First, observe that x^3+10 and $x^3+22.8$ both have the same derivative. In fact, we can really add any constant we want to any antiderivative for it will disappear when we differentiate. Hence, our equation $\int x^n\, dx = \frac{x^{n+1}}{n+1}$ should read

$$\int x^n\, dx = \frac{x^{n+1}}{n+1} + k,\text{ where k is any constant.}$$

Second, there is one special exponent that causes a problem, $n=-1$. If you try to apply our rule to $x^{-1}=\frac{1}{x}$, you will quickly find that no amount of differentiation of polynomials will yield $\frac{1}{x}$. In fact, recalling our three elementary derivatives above, only $y=\ln x$ has this derivative. Thus we can now give a general rule for our first two elementary antiderivatives:

$$\int x^n\, dx = \frac{x^{n+1}}{n+1} + k,\ n \neq -1,\text{ and }\int \frac{1}{x}\, dx = \ln(x) + k.$$

Example 3:

The following table gives the observed marginal revenues for different quantities sold over a period of time. Give the revenue function and find the revenue for 200 items.

quantity	0 - 30	31+
marginal revenue	$2	1.85

Since the marginal revenue is the derivative of the revenue function, we need the antiderivative. Since the marginal revenue function has different values reflecting a volume discount, the revenue function will have a jump discontinuity at q=31. Since the functions 2q and 1.85q have the required derivatives, our revenue function will have to be $R(q) = \begin{cases} 2q & 0 \leq q \leq 30 \\ 1.85q & 31 \leq q \end{cases}$. Note that since we are dealing with a revenue function, where it is known that if you sell nothing you make nothing (R(0)=0), we do not need to worry about the possibility that when the revenue function was differentiated some non-zero constant disappeared. Thus the revenue for 200 items is $R(200) = 2 * 30 + 1.85 * 170$.

Example 4: Antiderivative of e^x

Finally, our favorite derivative $y=e^x$ yields the immediate result that $\int e^x dx = e^x + k$. Thus, we have quickly derived the three elementary antiderivative formulas.

The three elementary antiderivatives:

$\int x^n dx = \frac{x^{n+1}}{n+1} + k, n \neq -1,$

$\int \frac{1}{x} dx = \ln(x) + k$

$\int e^x dx = e^x + k$

Finally, we need to combine these three elementary antidifferentiation rules with the appropriate theorems to see how much we can accomplish. To do this, let's first review the theorems we had on differentiation:

1. $[cf]' = cf'$ (take a constant outside to differentiate)

2. $[f \pm g]' = f' \pm g'$ (differentiate sums and differences a term at a time)

3. $[fg]' = f'g + fg'$ (The product rule)

4. $\left[\dfrac{f}{g}\right]' = \dfrac{gf' - fg'}{g^2}$ (The quotient rule)

5. $[f(g(x))]' = f'(g(x)) * g'(x)$ (The chain rule).

You may recall how you groaned when you first learned the product rule, the quotient rule, and the chain rule. Well, I have good news and bad news for you. First the good news: Theorems 3, 4, and 5 in general don't work in integral calculus. Just think, you don't have to learn what might be the horrible equivalents of these theorems.

Now for the bad news. Theorems 3, 4, and 5 have no general corresponding theorems in integral calculus that work. How can this be bad news? If you will recall, using the above 5 theorems we can work any differential calculus problem we get. Since we will no longer have these theorems, we will generally be unable to work many integral calculus problem we get. It is easy to see that the first two,

1. $[cf]' = cf'$ (take a constant outside to differentiate)
2. $[f \pm g]' = f' \pm g'$ differentiate sums and differences a term at a time)

have an exactly corresponding theorem in integral calculus:

1. $\int cf\,dx = c \int f\,dx$, ($\int 3x^2\,dx = 3 \int x^2\,dx$), and
2. $\int [f \pm g]\,dx = \int f\,dx \pm \int g\,dx$ ($\int (x^2 + x^3)\,dx = \int x^2\,dx + \int x^3\,dx$)

That, unfortunately, is about the end of the good news. We can come up with a rough equivalent of the product theorem (called "integration by parts"), but it does not always work, and when it does work, it can be time consuming and difficult to use. To show you how bad things can get very quickly, consider the function e^{-x^2}. This is a very common and useful function. Those of you who have had any statistics will recognize this as the "normal" or "bell shaped" curve function. It is easy to differentiate using the chain rule; the derivative is $e^{-x^2}(-2x)$. Unfortunately, we rarely want to differentiate it, but we nearly always want to antidifferentiate it. What is the antiderivative? There isn't one. Note that I did not say we couldn't find it; I said there isn't one (period).

How often does such a situation occur? Much more often than the standard calculus book would lead you to believe. Keeping this in mind, we will look at alternative ways of finding answers to questions requiring antiderivatives and the conceptual ideas required to do this correctly as we proceed.

Example 5:

Find the indefinite integral

a. $\int x^4 \, dx$ b. $\int \frac{1}{x^3} \, dx$ c. $\int \sqrt[3]{x^2} \, dx$ d. $\int \frac{2}{x^4} \, dx$

Solution:

a. $\int x^4 \, dx = \frac{x^{4+1}}{4+1} + k = \frac{x^5}{5} + k$

b. Rewrite before integrating: $\int \frac{1}{x^3} \, dx = \int x^{-3} \, dx = \frac{x^{-3+1}}{-3+1} = \frac{x^{-2}}{-2} = -\frac{1}{2x^2} + k$

c. Rewrite before integrating: $\int \sqrt[3]{x^2} \, dx = \int x^{\frac{2}{3}} \, dx = \frac{x^{\frac{2}{3}+1}}{\frac{2}{3}+1} = \frac{x^{\frac{5}{3}}}{\frac{5}{3}} = \frac{3}{5} x^{\frac{5}{3}} + k$

d. Rewrite before integrating: $\int \frac{2}{x^4} \, dx = \int 2x^{-4} \, dx = 2 \int x^{-4} \, dx = 2 \frac{x^{-4+1}}{-4+1} = 2(\frac{x^{-3}}{-3}) = \frac{-2}{3x^3} + k$

Example 6:

Find the indefinite integral:
a. $\int 5 \, dx$ b. $\int (x+4) \, dx$ c. $\int (3x^2 - 5x + 9) \, dx$ d. $\int \left(\frac{3}{\sqrt[4]{x^3}} + e^x + \frac{4}{x} \right) dx$

Solution:

a. $\int 5 \, dx = 5 \int x^0 dx = 5 \, x^{0+1}/(0+1) = 5x + k$

b. $\int (x+4) \, dx = \frac{x^2}{2} + 4x + k$

c. $\int (3x^2 - 5x + 9) \, dx = x^3 - (5/2) x^2 + 9x + k$

d. $\int \left(\dfrac{3}{\sqrt[4]{x^3}} + e^x + \dfrac{4}{x} \right) dx =$

$\int (3x^{\frac{-3}{4}} + e^x + 4x^{-1}) \, dx = 3 \dfrac{x^{\frac{-3}{4}+1}}{\frac{-3}{4}+1} + e^x + 4 \ln x =$

$3 \dfrac{x^{\frac{1}{4}}}{\frac{1}{4}} + e^x + 4 \ln x = 12x^{\frac{1}{4}} + e^x + 4 \ln x + k$

Example 7:

The marginal cost for producing q units of a product is modeled by MC = 28 - .03x and it costs $65 to produce one unit. Find:

a. the cost function.
b. the cost of producing 500 units.

Solution:

a. $C = \int (28 - .03q) \, dq = 28q - .015q^2 + k$

To solve for k, we know that C(1) = 65; therefore 65 = 28(1) - .015(1²) + k.

Now solve for k to get: k = 37.015

The Cost function is: C(q) = 28q - .015q² + 37.015

b. C(500) = 28(500) - .015(500)² + 37.015 = $8787.02

Information needed to work exercise 6.1

1. Elementary antiderivatives:

$\int x^n \, dx = \dfrac{x^{n+1}}{n+1} + k, \quad n \neq -1,$

$\int \dfrac{1}{x} \, dx = \ln(x) + k$

$\int e^x \, dx = e^x + k$

2. Elementary antiderivative theorems:

$\int cf \, dx = c \int f \, dx$, and

$\int [f \pm g] \, dx = \int f \, dx \pm \int g \, dx$

Exercise 6.1

1. $\int x^2 \, dx$
2. $\int x^5 \, dx$
3. $\int \dfrac{1}{x^2} \, dx$
4. $\int \dfrac{1}{x^4} \, dx$
5. $\int \sqrt[5]{x^3} \, dx$
6. $\int \sqrt[7]{x^2} \, dx$
7. $\int \dfrac{1}{\sqrt[5]{x^3}} \, dx$
8. $\int \dfrac{1}{\sqrt[7]{x^2}} \, dx$
9. $\int \dfrac{4}{x^3} \, dx$
10. $\int \dfrac{2}{x^5} \, dx$
11. $\int 6 \, dx$
12. $\int 4 \, dx$
13. $\int (x+3) \, dx$
14. $\int (2x-5) \, dx$
15. $\int (4x^3 - 6x^2 + 5x - 10) \, dx$
16. $\int (5x^4 - 4x^2 + 7) \, dx$

17. $\int (3\sqrt{x} - \frac{3}{x^2})dx$

18. $\int (5\sqrt{x} + \frac{4}{x^3})dx$

19. $\int -3e^x dx$

20. $\int 5e^x dx$

21. $\int 12x^{-1} dx$

22. $\int -\frac{5}{x} dx$

23. The marginal cost at a level of production of x items is MC=2x³+5x–6. The fixed cost is $750. Find the cost function.

24. Suppose a company has found that the marginal cost at a level of production of x thousand units is given by MC = $\frac{60}{\sqrt{x}}$ and the fixed cost is $24,000. Find the cost function.

25. The marginal cost is given by MC = x² – 2x + 3 and the cost of 3 units is $15. Find the cost function.

26. The marginal profit from the sale of x hundred items is MP = 4 – 6x + 3x² and the profit on 0 items is –$50. Find the profit function.

27. Find the revenue and the demand functions given the MR = 100 – 6x – 2x². Use the fact that R = 0 when x = 0.

28. A company produces a product for which the marginal cost is given by MC=2x–10 and the fixed costs are $140.

    a. Find the cost function.
    b. Find the average cost function.
    c. Find the cost of producing 100 units.

© 1995 Saunders College Publishing

Section 6.2 Area and the Fundamental Theorem of Calculus

Example 1: Oil Problem

Several years ago your brash and kind of crazy brother approached you for some capital to start drilling an oil well. You agreed to give him several thousand dollars of your hard-earned money for some stock in the well. He told you that there was "no possibility of failure since he could smell the oil". You decided to purchase much of the stock in the names of your children and wife (not entirely with her approval). Much to your wife's (and somewhat to your) amazement, he did strike oil. After your initial burst of joy had passed, the realities of being an owner in an oil well began to sink in. Tax forms. Instability in the oil market. More investments ("Dear, when are we ever actually going to see any of this money??"). After your first two rounds of completing your tax forms was finished you decided that you might like to get out of the oil business and reap the rewards of your wise investment. Sell your shares. But what is a fair price? Whatever this price is clearly must be attached to the volume of oil one can expect during the lifetime of the well and some wild guess as to the future value of oil. How can you get some idea of the volume of oil remaining? Let's analyze some of the data collected on your well.

You have records for how much oil has been pumped during the first 2+ years (25 months) of operation. Since pumping is determined by the market price and will fluctuate with this price, and since you wish to know how much it would pump if maximum flow were continued, you only use results during periods of maximum pumping operations. How much oil can this well be expected to pump during its lifetime given that it will be sealed when production drops below 2 barrels per month?

Month	1	2	3	7	9	12	13	18	25
Barrels	290	281	271	238	222	201	195	165	130

Two approaches to the solution of this problem are possible. They both require the same beginning. Specifically, some way has to be found to predict the long-term results. Our preparation to this point makes this a fairly easy problem for us.

Graphing the given points in the STAT editor, we recognize this as an ae^{-bx} type curve.

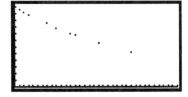 

Fitting an **exponential** curve (y=ab^x) to this data and **changing to a base "e"** exponential (if y=ab^x then y=ae^(x ln b)) we find that the desired function is (roughly)

$$y = 300 \, e^{-.033367 \, x} \doteq 300 \, e^{-\frac{x}{30}}$$

Note: This function is really only defined at the values observed at the end of each month, i.e. x=1,2,3,....

We are now faced with two remaining questions:

1. If pumping were maintained at maximum rate, at what time would production fall below 2 barrels per month, and
2. How many barrels could be pumped in total?

The first question requires us to solve the equation $300e^{-\frac{x}{30}} = 2$. Taking the log of both sides and solving we get n = 30ln(150) ≐ 150. In other words, if pumping were maintained at full capacity, the well would be depleted in 150 months.

Next, how much oil could be pumped during this time period? One method we can use to get the solution is to use our calculators to evaluate the function for each **integer** from 1 to 150 and then add the results. This process can be made simple by using the **sum** and the **seq** feature of the **MATH, MISC** menu.

To do this, enter the following on the home screen of your calculator. The entries made after the function are: variable used, beginning integer, ending integer, with an increment of 1.

Doing this yields 8,791 barrels.

Since we wish to introduce the ideas of integral calculus, let's consider another method of solving this problem. This method requires us to view the function we have derived in a slightly different fashion. If you will think about the function for a minute, you will realize that this function represents a rate: barrels per month. When we put in a value for x in the function $300e^{-\frac{x}{30}}$, the resulting number is the number of barrels produced during that month.

One can also view this function as one which gives the rate at any particular time. The trouble with this is that the function was derived based on output at the end of each month. If we wanted to find say the production for half of the third month, putting in 3.5 will not yield a reasonable answer without some interpretation. Just putting in 3.5 would yield the result roughly achieved from the middle of the second month to the middle of the third month, again, a one month span since that is how we derived the function in the first place. However, we can repair the situation as follows.

Suppose we wanted just the amount produced during the first two weeks of the third month. To get this we would calculate f(3.5) (the amount for the 4 week period from 2.5 to 3.5) and take 1/2 of the result. Thus, f(3.5)/2 yields roughly the result for he first two weeks of month 3. What if we wanted the result for just the first week? A good approximation would be given by calculating f(3.25)/4. Continuing in this fashion you can see that we could theoretically get a daily production estimation by calculating f(3 1/30)/1/30.

A convenient way to view these results is as follows. In the first case, since our function is defined one month at a time, we may also view the calculated volume as the amount of area in the accompanying picture.

OIL PRODUCTION BY THE MONTH
INTERPRETED AS AREA UNDER A CURVE

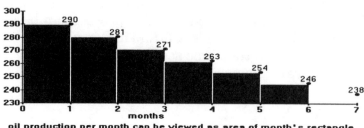

oil production per month can be viewed as area of month's rectangle
since the base=1 and the height is 1 month's production

Using this area idea, we can calculate the volume of production for every two weeks by calculating the functional values f(.5), f(1), f(1.5), f(2), etc. and selecting bases for our rectangles which are 1/2 of a month. This yields the following graph.

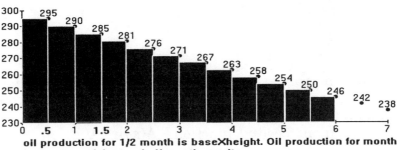

We can continue in this fashion, predicting output on a daily basis. Doing this for the life of the well would yield:

150(months)*30(days in a month)=4500 rectangles.

This could be done as before using **sum seq** but a knowledge of calculus yields a much faster result. Consider our prediction function again, but view it as a continuous function $300e^{-\frac{x}{30}}$ defined at every single real value in any interval. If we only had some **exact** method of finding the area under this curve, we could give the output of the well for any time interval we choose. How does one go about finding the exact area under the curve? Note that this curve is not one of those for which we have a formula (like rectangles and circles).

This brings us to an approach which provides great enjoyment for mathematicians. You may find that you will also enjoy this technique! We begin by supposing there is a function which gives the area under the curve starting at time 0. Never mind that we have no idea what this area function is. We will proceed by trying to find out how it would behave if there were one. In everyday terminology, this is the "if it walks like a skunk, eats like a skunk, and smells like a skunk, then it must be a skunk" approach.

What we want to find is a function that will give us the area between the curve and the x-axis. We want our function to give us this area between any two points quickly and easily by only requiring us to do some simple numerical evaluation. Let's give a name to our wishful area function: A(x). We now start making observations about A(x).

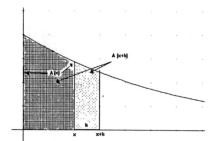

1. A(0)=0. There is no area at the start since we haven't pumped any oil.
2. If the function f has only positive values, A(x+1)≥A(x), since you will sweep out more area as you move any number of units to the right.
3. For any positive h and positive function f, A(x+h)≥A(x).

In order to work towards this solution we must make the following observations:

1. To find an approximation of the area between the curve f and the horizontal axis on an interval [v,w] one can fit a collection of rectangles with bases on the horizontal axis and one corner on the curve.
2. The approximation generally improves when you use more rectangles.
3. If you can figure out the general method to calculate the area of one rectangle you can use this technique to find the area of any (or all) of them.
4. The more rectangles you insert in the interval [v,w], the narrower each one has to be.
5. A rectangle with a corner on the curve does not usually "fit" correctly. The function would have to be a horizontal line for this fit to be correct.

Using these ideas we come up with the following diagram. Study it carefully. We wish to calculate **only** the area of the region bounded on the bottom by the x-axis, on each side by the values x and x+h, and on the top by the curve f(x). Note that the correct area is indicated, as well as the areas of two rectangles, the larger one enclosing the desired area, and the smaller one being enclosed by the desired area.

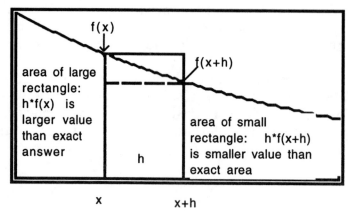

From looking at our figure we can make additional conclusions about the area function A(x):

1. A(x+h) − A(x) is the correct area in our diagram, since this is the area out to x+h minus the area out to x.
2. The area of the big rectangle is h*f(x) ≥ correct area, A(x+h) − A(x). That is, the big rectangle encloses all the area in question.
3. A(x+h) − A(x), the correct area ≥ h*f(x+h), the area of the small rectangle. That is, the correct area encloses all the area in the smaller rectangle.

The following relation summarizes these observation:

$$h*f(x) \geq \text{correct area} \geq h*f(x+h)$$
OR
$$h*f(x) \geq A(x+h) - A(x) \geq h*f(x+h)$$

Since h is positive, h>0, we can divide all inequalities by h giving the following:

$$f(x) \geq \frac{A(x+h) - A(x)}{h} \geq f(x+h)$$

Suppose now that we take the $\lim_{h \to 0}$ of the entire inequality expression:

$$\lim_{h \to 0} [f(x) \geq \frac{A(x+h) - A(x)}{h} \geq f(x+h)]$$

The $\lim_{h \to 0} f(x) = f(x)$, the $\lim_{h \to 0} f(x+h) = f(x)$, so we immediately get:

$$f(x) \geq \frac{A(x+h) - A(x)}{h} \geq f(x)$$

We have just derived another property of the area function! Since the expression in the middle is both greater than or equal to f(x) and less than or equal to f(x), it must be f(x). Thus we have:

$$\lim_{h \to 0} \frac{A(x+h) - A(x)}{h} = f(x)$$

But what good is that? What is the value of $\lim_{h \to 0} \frac{A(x+h) - A(x)}{h} = f(x)$? If you will recall our earlier work in derivatives, you will realize that the limit of this fraction is the derivative of the function in question. Thus, we have:

$$\lim_{h \to 0} \frac{A(x+h) - A(x)}{h} = A'(x) = f(x)$$

In the case in question, our function f(x) is $300e^{-\frac{x}{30}}$. **IF** there is an area function A(x), its derivative A'(x) is $300e^{-\frac{x}{30}}$. The question now resolves to this (not so easy) task.

Does there exist a function A(x) such that when you differentiate it you get $300e^{-\frac{x}{30}}$? For the moment we will not tell you how we got it, but the answer is yes and it is $A(x) = -9000e^{-\frac{x}{30}}$. You can check this result by differentiating A(x):

$$A'(x) = \left[-9000e^{-\frac{x}{30}}\right]' = 300e^{-\frac{x}{30}}.$$

For obvious reasons, $-9000e^{-\frac{x}{30}}$ is called the **antiderivative** of $300e^{-\frac{x}{30}}$.

This yields a very important theorem:

FUNDAMENTAL THEOREM OF CALCULUS:

The area between the positive function f(x) and the x-axis between any two points v and w where v<w can be found by calculating the antiderivative of f(x), say A(x), and evaluating it thus: A(w)-A(v). This is usually written as follows:

$$\int_v^w f(x)\,dx = A(w) - A(v)$$

where A(x) is the antiderivative of f(x), and the "dx" tacked on the end indicates the variable being dealt with is x.

Returning to the problem in question, to find an approximation to the total amount of oil pumped during the 150 month period, we calculate the antiderivative and the difference A(150)-A(0). Your calculator will now perform the following calculations:

$$\int_0^{150} 300e^{-\frac{x}{30}}\,dx = -9000e^{-\frac{x}{30}}\Big|_0^{150} = -9000e^{-\frac{150}{30}} - (-9000e^{-\frac{0}{30}}) = 8939.33$$

We can use the GRAPH editor or we can use the CALC, fnINT of our calculator to evaluate this antiderivative from 0 to 150. When using the Graph editor, press MATH, ∫f(x). We must now use the arrow buttons to enter the bounds "0" and "150", pressing ENTER after each. Our results:

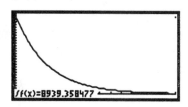

Using CALC, fnINT, we enter the information as follows:

```
fnInt(y1,x,0,150)
         8939.35847701




 evalF  nDer  der1  der2  fnInt
```

Note: y1 is the name of the function in the graph editor. We can also type the function $300e^{-\frac{x}{30}}$ in its place.

You might note that there is a slight discrepancy between this answer, 8939, and the **sum seq** answer of 8791. First, given the question we want answered, both answers are equally acceptable since they are only approximations. Second, the discrepancy is caused because we pick up a little too much area each month using the **antiderivative** approach. This can easily be solved by moving the starting and ending points, which we will do when it becomes important.

Such a deadly accurate solution is not important in this problem and would, in fact, interfere with our desire to get a quick answer. You might note after all this discussion and explanation that, after you learn what you are doing, the solution to this problem will require only slightly more than the time it takes to type data into your calculator. In our case, from the end of typing in the data to the final solution took a little less than 1 minute. This is pretty impressive for a problem of this magnitude and potential difficulty.

In addition, you should observe the following. Using this new technique not only can we find the amount of oil pumped in the time period 0 to 150, we can also find the maximum amount of oil that can be pumped in any time period. For example, the area under the curve from 40 to 60 will give the approximate maximum amount that could be pumped from the 40th through the 60th month. This amount is:

$$\int_{40}^{60} 300e^{-\frac{x}{30}}\, dx = -9000e^{-\frac{x}{30}}\Big|_{40}^{60} = -9000e^{-\frac{60}{30}} - (-9000e^{-\frac{40}{30}}) = 1154.35$$

A summary of our new ideas

To get a little more general picture of what we have just done, let's take another look at our results. We have really drawn two conclusions:

1. Given a curve f(x), the area between the curve and the horizontal axis represents the solution to certain types of problems.
2. For positive functions, the area between the curve and the x-axis can be found by finding the antiderivative and taking the difference of the values at the end points.

© 1995 Saunders College Publishing

The enormity of these discoveries should not be underestimated. Don't let the fact that in **this particular application** of our new concept, absolute accuracy was unimportant and our **sum seq** answer would have been sufficient. In applications we shall do shortly, the exactness of the calculus result will be required. Absolute accuracy (or as nearly absolute as we can get) in most problems which occur in engineering and the sciences is required. The list of applications of integral calculus is nearly endless. Our concern with those that have applications to business and economics will force us to leave out a multitude of applications in other areas.

Finally, no introduction to integral calculus would be complete without pointing out that finding antiderivatives varies from tedious, to extremely difficult, to impossible, <u>very quickly</u>. This possible-impossible dichotomy will be addressed as we move through our discussion. Over the centuries, much time has been devoted to finding antiderivatives. It would take us several years to cover all the ingenious techniques developed for finding antiderivatives. The reason for this tremendous effort at finding antiderivatives is two-fold.

First, using rectangles, any number (short of infinity) yields only an approximate answer. Finding the area function is the only way to get the "correct" answer.

Second, without high speed computing equipment, using rectangles or other numerical techniques leaves you with mountains of calculations. Such an error prone approach gave considerable motivation for finding antiderivatives whenever possible, no matter how difficult the task.

Fortunately for you, much of this motivation has been removed since you do have high speed computing equipment available. Thus, in our development, we will usually take an approach which is somewhat different than the classical approach found in most texts. Whenever it is convenient, we will use antiderivatives (you just can't beat perfection!). Otherwise, we will use a numerical technique. Don't get the idea that using a numerical technique will be the solution to all your problems. There are several dangers in using a numerical technique:

1. Some problems do not have a finite solution, that is the answer is infinitely large. Unfortunately, a calculator, being the finite calculating machine that it is, will still give a finite answer. To merely say this is misleading would be a gross understatement.
2. Some problems having a perfectly reasonable solution cannot be solved by the numerical technique you have chosen because of round-off problems. Imagine that the solution to your problem involves adding .01 to 1,000,000,000. a billion times in a machine that can only store 10 digits. Since there is no room to store 1,000,000,000.01, and .01 is tiny compared to the original number, the calculator is designed to discard the smaller number when a choice has to be made. Hence, in a blink, the calculator discards
1,000,000,000*.01=10,000,000. OUCH.

Hence, another consideration we will have as we move along is trying to determine how far off our answer might be from the correct answer when using a numerical technique. In the oil well problem we have just worked and in many types of business problems, either method yields a solution. The antiderivative technique yields the exact solution with most any numerical technique yielding a very close solution.

Exercise 6.2

1. Use the fnINT of the CALC editor to evaluate the following:

   a. $\int_0^1 3 \, dx$ b. $\int_2^5 3x \, dx$ c. $\int_1^3 (5x-2) \, dx$

   d. $\int_2^6 e^{3x} \, dx$ e. $\int_0^{10} e^{-x} \, dx$ f. $\int_0^4 \frac{2}{\sqrt{2x+3}} \, dx$

2. Graph $y = 4x^2 + 2$ and use the $\int f(x)$ of the MATH menu of the GRAPH editor to find the are bounded by the function, the x-axis, between x=0 and x=4.

3. When a certain text book is published the monthly sales rate in 1,000's R(t) is given roughly by the function:

$$R(t) = 100 \frac{t^2 e^{-\frac{t^2}{2}}}{3!}, \; t > 0, \; t \text{ in years.}$$

Since this equation is the sales rate, it is the derivative of the sales. Thus, we can find total sales for the first 7 months by calculating

$$\int_0^{\frac{7}{12}} 100 \frac{t^2 e^{-\frac{t^2}{2}}}{3!} \, dt$$

Calculate this answer.

4. Find the Capital Accumulation over a t-year period, given the rate of investment dI/dt, by evaluating the definite integral

$\int_0^t (dI/dt)\, dt$ for the follwoing:

    a. dI/dt = 800   6 year period

    b. dI/dt = 200t 4 year period

5. Find the increase in the cost of increasing production from x=100 to x=105 if the MC = 3x +5.

6. Find the increase in the revenue if production is increased from x=15 to x=20, when MR = 50 – 2x.

7. A company has found that its expenditure rate per day (in thousands of dollars) of a particular job is given by E(x) = 6x + 5, where x is the number of day since the start of the job.

    a. Find the total expenditure for the first 8 days of the job.
    b. Find the total expenditure for the job if it takes 2 weeks to finish the job.

Section 6.3 Probability Density Functions

One of the most frequently asked questions concerns using mathematics to make predictions. Obviously, an accurate prediction can save much money and time. We have done this in many different (and seemingly unrelated) situations. We now consider another brought about by a student's question. One of my current students works in a shoe store. They are having to unload a fair number of size 5 women's shoes at considerably under their cost just to get them out of the store. Is there a way to avoid this problem when ordering shoes? The answer is an emphatic **yes**.

Since so many shoes are sold each year, complete data about shoes is easy to come by and well known. People's feet in general follow a normal distribution given by the curve $y = ae^{-b(x-c)^2}$. You have done enough work to know that given some data points you can evaluate a, b, and c in this equation and fit a curve. However, this is not really necessary for data such as shoes where nearly everything you would want to know is already known, or for a curve such as the normal distribution which has been studied by statisticians for centuries.

In this particular case, it is a well known fact that the distribution of sizes follows this curve:

$$y = ae^{-b(x-c)^2} = \frac{1}{\sqrt{2\pi * sd^2}} e^{-\frac{1}{2*sd^2}(x-mean)^2}$$

where the mean or average of women's shoe sizes is known to be about $7\frac{1}{2}$ and "sd" is the standard deviation which is known to be about 1. While all of you will understand the meaning of "mean" or "average", many of you will not understand the meaning of "distribution" and "standard deviation" so we will digress from the problem for a moment to explain those terms.

Distribution

Many years ago (1962 to be exact) W. J. Youden, a statistician from the Bureau of Standards in Washington was asked to explain the meaning of "distribution" to a bunch of high school students. His method was so effective that I will use it here. He bought 100 empty flower pots and had the students plant one seed in each pot. These were then left in a greenhouse and cared for 1 month. At the end of this time the smallest plant was placed against a wall at one end of the greenhouse and the largest at the other end. In between there was room for 15 more pots. From the remaining pots the students choose 15 of increasing size establishing a scale consisting of 17 potted plants of increasing size along the wall. All of the remaining 83 plants were now stacked in the column that most nearly approximated their size. When they were done the students were surprised to find that the plants were arranged in the following pattern:

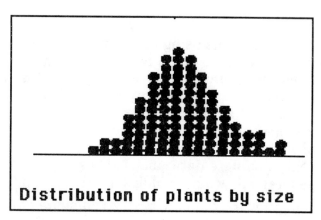

The arrangement of these pots is called the "distribution" of the plants by size. Note the shape of the distribution. It roughly follows the bell shaped or Normal curve. People are like plants. We are distributed just like this in many respects, height and shoe size being two of them. Since we know this, we need only know where the mean or average occurs, and how much the distribution is spread out. This spreading of the distribution is called the standard deviation.

Standard Deviation

The standard deviation of a distribution is a measure of how much your data is spread out from the mean or average. Look at the accompanying graphs.

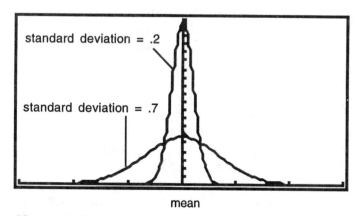

Normal Curves

In both graphs the mean occurs at 0. In one case we have a normal distribution which has a standard deviation of 0.2 and, in the other case, a standard deviation of 0.7. Note how the distribution for the standard deviation of 0.2 is gathered closely around the mean. In the other case, note how the data is spread out from the mean considerably more.

You should make a mental note here that the less the data is spread out, the better your predictions about the data will be. Hence, a small standard deviation is much better than a large one. Be a little careful how you interpret the term small standard deviation. When weighing elephants, a small standard deviation is 100 lbs. When measuring machine tool parts, .1 inch is large, one would wish for 0.0001 inch.

In the case of women's shoes, the mean is about $7\frac{1}{2}$ and the standard deviation is about 1. Thus, our normal curve $\frac{1}{\sqrt{2\pi * sd^2}} e^{-\frac{1}{2*sd^2}(x-mean)^2}$ reduces to $\frac{1}{\sqrt{2\pi}} e^{-\frac{1}{2}(x-7.5)^2}$.

How does one use this curve to buy shoes? You may have noted the peculiar fraction, $\frac{1}{\sqrt{2\pi * sd^2}}$, in the front of the equation. This fraction has been carefully chosen so that the total area under the curve adds up to 1. When the total area under a curve adds up to 1, each portion of the area under the curve may be considered to be the probability of finding someone or something in that range. Statisticians have a special name for such curves, they call them **probability density functions**. Any probability density function will have the peculiar characteristic that the total area between it and the horizontal axis totals to 1. This permits there to be a direct connection between the area under a portion of the curve and the probability of something in that range occurring. For example, one would not ask the probability of getting exactly 5, one would ask the probability of finding something in the range 4 to 6, or $4\frac{1}{2}$ to $5\frac{1}{2}$.

Returning to our shoe problem, look at the accompanying graph. This graph shows exactly what the distribution of adult women's feet look like. This curve has the special multiplicative factor applied so the total area under the curve is 1. Note that in general this area extends from -∞ to ∞ on the x-axis.

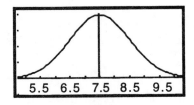

mean

Normal Curve for Women's Shoes

In our graph, suppose the area from 6 to 8 were found to be 0.6. This would mean that if 100 women were to walk into your store you would expect 60 of them to have shoe sizes between 6 and 8. Suppose the area from 4 to 6 was found to be 0.07. Again, out of 100 women you would expect only 7 of them to have this shoe size.

You should now begin to see how we can use this idea to order shoes. We will first find the probability of a certain size occurring, then, using this number, we will multiply by the total number of customers we expect or by the total number of shoes we need to buy.

First, we have a few other considerations. Since we want to know how many of a particular size, say 7, that we need to buy, we need to know this probability. However, there is no area directly over the number 7. You need to realize that there is also no **exact** size 7. To find the probability of a size 7 person walking into our store, we need to calculate the area from $6\frac{3}{4}$ to $7\frac{1}{4}$ since these foot sizes would most likely be placed into a size 7 shoe.

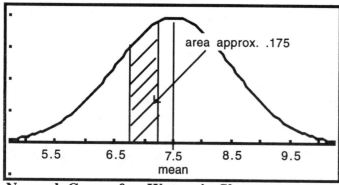

Normal Curve for Women's Shoes

This area is calculated by evaluating $\displaystyle\int_{6.75}^{7.25} \frac{1}{\sqrt{2\pi}} e^{-\frac{1}{2}(x-7.5)^2} = 0.175$

You would interpret this number by saying that if 1000 people entered your shop, about 175 would want a size 7 shoe or, if you were ordering 1000 pairs of shoes, 175 should be size 7.

Hence, we now see how to calculate the probability for each given shoe size: one has to go $\frac{1}{4}$ above and $\frac{1}{4}$ below the given size and then find this area under the curve. Also, it should be clear to you how this information can be used to order shoes. For example, if you decided to order 100 pairs of shoes, how many size sevens should you order? Since the probability is 0.175 you would order either 17 or 18 pairs out of the 100 pairs.

There is, however, a fly in the ointment (isn't there always?). Shoes not only come in **lengths** but also **widths**. Widths range from AAAAA (very, very, very, very narrow) to around EEEE (very, very, very wide). Let's take our example of the size 7 shoe and see how many of each width we should order. Remember we concluded that we would need to order about 17 or 18 pairs of this size.

As you may have already guessed, shoe widths are also normally distributed. The mean for women's shoe widths is about "B" and the standard deviation is about 1 letter. Hence, if we arbitrarily assign the number 0 to B, we can put in mean=0 in our formula (makes it lots easier to write out) and then the letter widths A=1, AA=2, C=-1, D=-2, etc. Thus, for our next calculation we will find the probability that someone walks in with a size 7B. This is essentially the question: What is the probability that someone with a size seven foot **and** a B width enters. We already know that the size 7 answer is 0.175, what is the B width answer and how do we combine the information?

Fortunately for us, statisticians have also studied this problem extensively and the conclusion is this:

Probability for independent events:

If two things occur which are independent of each other with probabilities x and y, the probability of both of them occurring is x*y.

We will assume that length of foot and width of foot are independent of one another so all we have to do is find the probability of a width of B and multiply our two answers.

Since we have another standard deviation of 1 and our mean=B=0, we need to find the area under the normal curve from -0.5 to .5 (by our number assignment, widths progress by whole numbers, ...-3,-2,-1,0,1,2,3... and not half sizes like lengths).

Substituting in our equation $\dfrac{1}{\sqrt{2\pi * sd^2}} e^{-\frac{1}{2*sd^2}(x-mean)^2}$ we get $\dfrac{1}{\sqrt{2\pi}} e^{-\frac{1}{2}(x-0)^2}$.

Calculating, $\int_{-.5}^{.5} \dfrac{1}{\sqrt{2\pi}} e^{-\frac{1}{2}x^2} = 0.383$. Thus the probability for someone with size 7B is 0.067. If you were ordering 100 pairs of shoes you would order 6.7 shoes (either 6 or 7) of size 7B.

Exercise 6.3

1. You own a small boutique shoe store. You decide to order 50 pairs of shoes in a particular style.
   a. How many pairs of size $7\frac{1}{2}$ B should you order?
   b. How many pairs of size $5\frac{1}{2}$ B should you order?

c. Write down any corrections you might need in your basic equation if you feel that this shoe will be a dress shoe primarily purchased by high school girls aged 14-17.
d. Now assume that one of your best customers wears size 8 AAA. Calculate the probability of an 8 AAA, calculate how many pairs of these shoes you **should** order in a shipment of size 50 and, finally, how many do you order and why?

2. Men's shoes have a mean length of about 10, a mean width of about C, and a standard deviations for length about 1 and width one letter. Write down the necessary normal equations for men's shoes.

3. A short time ago one of our students got a job at a general store in Idaho for the summer. Shortly after starting work the owner had a heart attack and was confined to his home. The U.S. Forest Service wanted the student to place their usual order for 300 pairs of men's boots. What adjustments will need to be made in length and width for the predictions to come out accurately? Make those adjustments and:

a. Calculate how many boots of size $10\frac{1}{2}$ EE you will need to order.

Some of you would have copped out and (heaven forbid) would have asked the owner how to order. We are pleased to report to you that our student did the necessary calculations and had to reorder 3 pairs. The owner had never done better than 20 and averaged 50 returns.

4. You need to order 1,000 men's sweaters. You know that the mean size is Large. Men's sweaters come in small, medium, large, x-large, and xx-large, in that order. Assume that the standard deviation between sizes is 1 unit. The equation is:

$$\frac{1}{\sqrt{2\pi * sd^2}} e^{-\frac{1}{2*sd^2}(x-mean)^2}$$

a. How many of each size should you order?
b. What colors to order is also a problem. Do you think color choice is normally distributed? Explain your answer. How **would** you figure out the answer to this question?

TI-85: CALCULATOR INSTRUCTIONS I

OBJECTIVES: The student should be able to enter and analyze statistical data.

This will include:
1. entering data points (one or more sets) in the STAT editor,
2. graphing, using DRAW, in the STAT editor to display a scatter plot(s) and an xyLINE graph(s),
3. performing regression analyses (curve fitting),
4. forecasting a statistical data value.

1. Entering data points in the STAT editor:

From the STAT menu select EDIT

a. Enter the name of the list of x values, then press ENTER. You may use the displayed name (xStat), select a name from the menu, or type a new name. This name is case-sensitive and may contain up to eight letters.
b. Enter the name of the list of y values the same way, then press ENTER.
c. If the lists are new, only the first data point is displayed. The x element is blank and the y element has a default of 1. If the lists already exist, the contents are displayed. If you used xStat and yStat for a previous list and you wish to use these names for new list, then press F5 (CLRxy) to clear old data before entering new data.

Example 1:

Enter the following data points in the STAT editor.

x	-1	0	2	5	8
y	3	-1	0	2	5

Press STAT, EDIT

a. We will use xStat for the name of the xlist. Press ENTER.
b. We will use yStat for the name of the ylist. Press ENTER then F5 to clear any previous data using these names.
c. Beside x_1 = enter -1 then press ENTER.
   Beside y_1 = enter 3 then press ENTER.
   Continue this process until all data points are entered.

APPENDIX A

Example 2:

Enter the following data points in the STAT editor.

x	-5	0	1	2	6
y	-1	1	3	2	-2

Press STAT, EDIT

a. Type the name of the xlist. The keyboard is set in ALPHA-lock; therefore all letters typed will be upper case. If you wish to use lower case letters, press 2nd then ALPHA. Let us name this xList as xStat2 for example two. You may leave the existing "xStat" by pressing the right arrow button. Now type "2" then press ENTER.
b. Follow the same procedure to name the ylist as yStat2.
c. Beside x_1 = enter -5 then press ENTER.
   Beside y_1 = enter -1 then press ENTER.
   Continue this process until all data points are entered.

2. **Graphing, using DRAW, in the STAT editor to display a scatter plot(s) and an xyLINE graph(s):**

   NOTE: The STAT DRAW editor is closely connected to the GRAPH editor.

   1. You must clear or deselect any functions in the GRAPH editor or they willbe plotted.
   2. The RANGE in the GRAPH editor will define the viewing window for the DRAW in the STAT editor.

   To display a scatter plot of the data points in Example 1:

   a. Press GRAPH
      1. Press F1 (y(x)). If there is a function(s) beside y1=, then either press CLEAR or F5 (SELCT) to deselect the function(s).
      2. Press 2nd F2 (RANGE). Since we will eventually draw Example 1 and Example 2 on the same window, we will now set the RANGE as follows:

         xMin = -6   yMin = -3
         xMax = 10   yMax = 8
         xScl = 1   yScl = 1

   b. Press STAT, EDIT.
   c. Press F1, ENTER, F2, ENTER.
      Recall the names of the xlist and the ylist for Example 1 were xStat and yStat.
   d. Now press 2nd F3 (DRAW) and clear any previously used drawings by pressing F5 (CLDRW).
   e. Press F2 (SCAT).

APPENDIX A

You should now see the following graph:

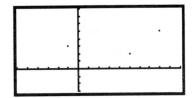

To display an xyLINE graph for Example 1: (Plots and connects data points with lines)

Press F3 (xyLINE). If the menu covers part of the graph, press the CLEAR button. Your graph should look like the following:

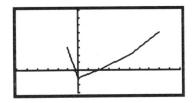

To display an xyLINE graph for both Example 1 and Example 2:

Press STAT, EDIT, xStat2, ENTER, yStat2, ENTER, 2nd F3 (DRAW), xyLINE. You now should have both xyLINE graphs on your screen:

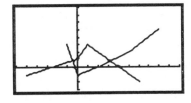

3. Performing regression analyses (curve fitting):

After selecting or entering the lists to use, select CALC from the STAT editor. Press MORE to move around the menu. Enter the statistical calculation of choice by pressing one of the menu keys. The analysis is calculated using the least-squares fit.

Key	Analysis	Regression Equation
F1(1-VAR)	One-variable results	y=integers ≥ 0
F2(LINR)	Linear regression	$y = a + bx$
F3(LNR)	Logarithmic regression	$y = a + b \ln(x)$, $x>0$
F4(EXPR)	Exponential regression	$y = ab^x$, $y > 0$
F5(PWRR)	Power regression	$y = ax^b$, $x > 0$, $y > 0$
MORE		
F1(P2REG)	2nd-order polynomial	$y = a_2 x^2 + a_1 x + a_0$
F2(P3REG)	3rd-order polynomial	$y = a_3 x^3 + a_2 x^2 + a_1 x + a_0$
F3(P4REG)	4th-order polynomial	$y = a_4 x^4 + \ldots + a_0$
F4(STREG)	stores the regression equation to a variable. Enter the name after **Name =.** You may use this key to store the regression to **y** to be used in the graphing editor. Remember to use a lower case y.	

For two-variable regression lists, except polynomial regression, **corr** (correlation coefficient) measures the goodness of the fit of the equation with the data points. In most cases, the closer **corr** is to 1 or -1, the better the fit. If **corr** = 0 then x and y are completely independent.

Example 3:

A company's annual sales for the first five years in business is given by the following table.

x	1	2	3	4	5
y	7	10	11	14	15

Find the equation of the least squares line. What is the correlation coefficient?

Step 1:   Enter data points in the STAT editor. Name **xStat3, yStat3**.
Step 2:   Press 2nd F1 (CALC), ENTER, ENTER. Press F2 (LINR) - Linear Regression. Your screen should read as follows:

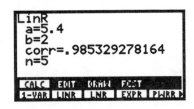

Step 3:   Equation - $y = 5.4 + 2x$ corr = .985329 (pretty good fit)

To display the regression curve in the STAT editor:

a. Press 2nd F3 (DRAW). Clear any existing graph by pressing F5 (CLDRW).
b. Press F4 (DRREG) - Draw Regression.
c. Press F2 (SCAT) to view the data points and the regression curve.

Example 4:

Draw the scatter plot and the regression curve for Example 3.

Step 1:   Clear or deselect any functions that are in the GRAPH editor.
Step 2:   Set an appropriate RANGE (viewing window) in the GRAPH editor.
Step 3:   Clear STAT DRAW window (F5) after pressing F3 (DRAW).
Step 4:   Press F2 and F4. Your screen should be similar to:

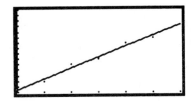

Example 5:

Perform regression analyses for the data points in Example 3 using:

a. Logarithmic Regression
b. Exponential Regression
c. Power Regression

Compare the correlation coefficients. Which regression of these and the linear, seems to be the "best fit"? Now draw the "best fit" curve and the scatter plot. Your window should be similar to the following:

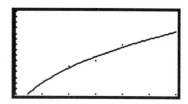

Power regression $y = ax^b$, $y = 7.0015x^{.47266}$ with corr. = .990697

4. Forecasting a statistical data value:

a. Press the FCST key (F4) in the STAT editor. Note: The current regression model is displayed on the top line.
b. Enter the x or y value beside x= or y= . Move the cursor, using arrow keys, beside the variable to be solved.
c. Press F5 (SOLVE).

Example 6:

Using the power regression in Example 5, forecast the sales for x=6. Your screen should be as follows:

TI-85: CALCULATOR INSTRUCTIONS II

OBJECTIVES: The student should be able to enter and graph functions.

This will include:

1. entering a function(s) in the GRAPH editor,
2. setting an appropriate viewing window,
3. displaying the graph,
4. exploring the graph with the TRACE feature,
5. exploring the graph with the ZOOM features,
6. analyzing the graph with the MATH features.

1. **Entering function(s) in the GRAPH editor:**
   a. Press the GRAPH key.
   b. Press the y(x) (F1) menu key.
   c. Type in the function(s) beside the appropriate y(x)(s), using the key **x-VAR** to represent the independent variable.
   d. Press ENTER after each function that is entered.

2. **Setting an appropriate viewing window:**
   Use the same procedure that we used when we set the viewing window for the DRAW in the STAT editor. Press the RANGE (F2) key.

3. **Displaying the graph:**
   Press the GRAPH (F5) menu key.

4. **Exploring the graph with the TRACE feature:**
   Press the TRACE (F4) menu key. The blinking cursor shows the location of the point given by the x and y coordinates at the bottom of the screen. Use the left or right arrow keys to move around the screen on the curve. Panning left or right off the display will automatically change the viewing window and update the xMin and xMax. While using the trace feature, pressing the ENTER key will adjust the viewing window so that the cursor will be in the center of the screen. The RANGE window will automatically be updated.

5. Exploring the graph with the ZOOM features:

There are fifteen ZOOM features. You can explore all of these on pages 4-18 through 4-22 of your calculator guidebook. I will cover most of these in class using your guidebook as a reference. The following is a list of the features that we will use with only a brief description:

Feature	Brief Description
BOX	Boxes in an area of the curve (cursor defined), then redraws the graph on a new screen - showing only the area in the box.
ZIN	Magnifies the graph centered around the cursor location.
ZOUT	Displays more of the graph.
ZSTD	Displays the graph using the default RANGE window.
ZPREV	Sets the RANGE window back to the settings of the previous window.
ZFIT	Sets yMin and yMax to include the min. and max. y values of the function between the xMin and xMax that you select.
ZOOMX	Adjusts the horizontal axis of the RANGE window. Use this feature (one or more times) to rotate a nearly horizontal curve away from the x-axis for a "nice" graph.
ZOOMY	Adjusts the vertical axis of the RANGE window. Use this feature (one or more times) to rotate a nearly vertical curve away from the y-axis.

6. Analyzing the graph with the GRAPH MATH features:

There are thirteen GRAPH MATH features. You can explore all of these on pages 4-24 through 4-28 of your calculator guidebook. I will cover most of these in class, as we need them, using your guidebook as a reference. The following is a list of the features with a brief description.

Feature	Description
LOWER	Lower bound of interval in question.
UPPER	Upper bound of interval in question.
ROOT	Finds the roots of a function in a given interval.
dy/dx	Finds derivative (slope) of a function at a given point.
∫ f(x)	Finds the definite integral of a function in a given interval.
FMIN	Finds the minimum of a function on a given interval.
FMAX	Finds the maximum of a function on a given interval.
INFLC	Finds inflection point of a function at a given point.
YICPT	Finds the y-intercept of a function.
ISECT	Finds the intersection of two functions in a given interval.
DIST	Finds the distance between two points on the display.
ARC	Finds the distance along a function between two points on the function.
TANLN	Draws the tangent line at a point.

Example 1:

Graph   a. **y = (3/5)x - 4** and b. **y = -2x + 6** on the same screen.
Find the following for both a and b by using the calculator features:

1. y-intercept
2. x-intercept
3. intersection of a and b

Solution:

1. Press GRAPH key.
2. Press y(x) key.
3. Beside "y1 =" type (3/5)x - 4 by pressing the following keys:
   (3 ÷ 5)x – VAR – 4 then ENTER. Beside "y2 =" type the other equation.

4. Press RANGE (F2) and use the default setting or press ZOOM (F3) then ZSTD (F4).
5. When using the ZSTD menu key, the graph will display immediately after pressing F4. If you do not use the ZSTD then press GRAPH (F5).

Your screen should be as follows:

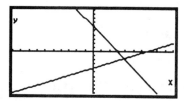

Now to use the calculator features to find:

1. the y-intercept for:

    graph a) - Press **MORE, MATH, MORE, YICPT, ENTER** and your screen should look as follows:

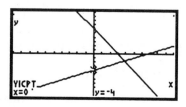

Press the **GRAPH** key

    graph b) - Press **MORE, MATH, MORE, YICPT**, then use the up arrow key to place the cursor on line b, then press **ENTER** and your screen should look as follows:

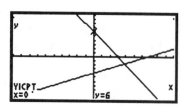

Press the **GRAPH** key, then **MORE**

2. the x-intercept for

   graph a) - Press **MATH**, **MORE**, **ROOT**, **ENTER** and your screen should look as follows:

   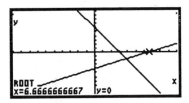

Press the **GRAPH** key, then **MORE**

   graph b) - Press MATH, MORE, ROOT, upper arrow key, ENTER and your screen should look as follows:

   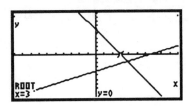

Press the **GRAPH** key, then **MORE**

3) the intersection of the lines

   Press **MATH, MORE, ISECT**, move arrow key to what looks like the intersect point, ENTER, ENTER and your screen should look as follows:

   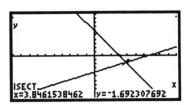

TI-85 CALCULATOR INSTRUCTIONS II-B

These are Examples 3 and 4 in section 2.1. This is a little different approach.

Example 3: Graph y=-110x+15

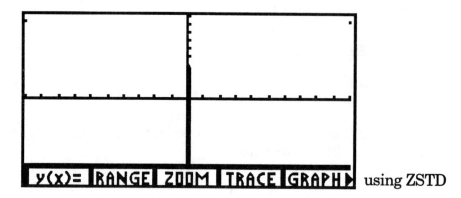 using ZSTD

Knowing that the y intercept is 15 and the x-intercept =15/110 which is < 1, we might

set our RANGE window to the following: .

We would then get .

Example 4: Graph $y = \frac{7}{327} x + \frac{23}{327}$

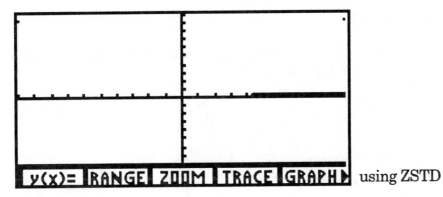 using ZSTD

Knowing that the y-intercept is very close to 0 (i. e. 23/327), manually change the Range Window to the following:

 to get the following graph 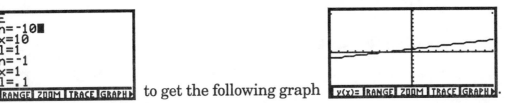.

Leaving the yScl = 1 instead of changing it to .1, merely removes the tick marks between -1 and 1 on the y-axis. The graph looks the same.

Using the ZOOMX feature (ZOOM, MORE, MORE, F2, ENTER) with our original (ZSTD) graph, we get the following graph:

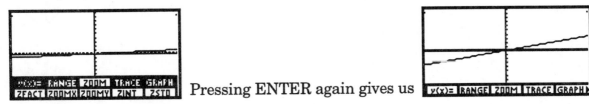 Pressing ENTER again gives us

APPENDIX A xiii TI-85: II-B

with the following RANGE window  set by the calculator.

At this point, we should still manually adjust this window for a better graph. We need to show more clearly the x and y intercept.

We might again use the following setting: to get the

following graph:

TI-85: CALCULATOR INSTRUCTIONS III
USING THE SOLVER

I. Solving Equations

To display the SOLVER equation entry screen, press SOLVER (2nd GRAPH). The solver displays the last equation used in the solver (if any). Press CLEAR if an equation is displayed beside "eqn:". Type in the equation to be solved.

Example 1:

To solve $12x - (4x - 6) = 3(9 - 2x)$ for x, type in the equation and then press **ENTER**. Clear any previous x value on the screen that is beside x=. Then press **SOLVE** (F5). You should see x = 1.5.

Example 2:

To solve $6x^2 + 11x + 4 = 0$ for x, type in the equation beside "eqn." after clearing any previous equation. Then press **ENTER**. Clear any previous x value on the screen that is beside x=. Then press **SOLVE** (F5).
You should see x = −.50000000000001.

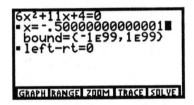

Since this is a quadratic equation, there may be at most one more solution. To explore the possibility of another solution, we might look at the graph or make a guess as to the other root. When making a guess, the calculator will give the root that is closest to that guess. To look at the graph of this quadratic function press **ZOOM, ZSTD**.

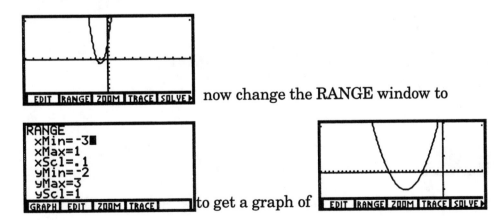

APPENDIX A   xv   TI - 85 III

Now we can use **TRACE** to move the cursor closer to the other root. Looks close to x=−1.3. Press **SOLVER** (2nd GRAPH), then **ENTER**. Now x is set at the last position of the cursor. Press **SOLVE**. The other root is x = −1.3333333333333.

This example would have been easier to do by using ROOT in the GRAPH editor.

II. Storing Formulas To Use In The SOLVER

To store formulas in the solver and then use the solver to solve for any variable in the equation, first enter the equation on the Home screen.

Example 1

On Home screen type in the following formulas pressing **ENTER** after each.

F1=P(1+(r/N))^(N*t)
F2=P*e^(r*t)

Now press **SOLVER** (2nd GRAPH). Clear anything beside"eqn:". Now press **RCL** (2nd STR->) **F1. ENTER, ENTER.**

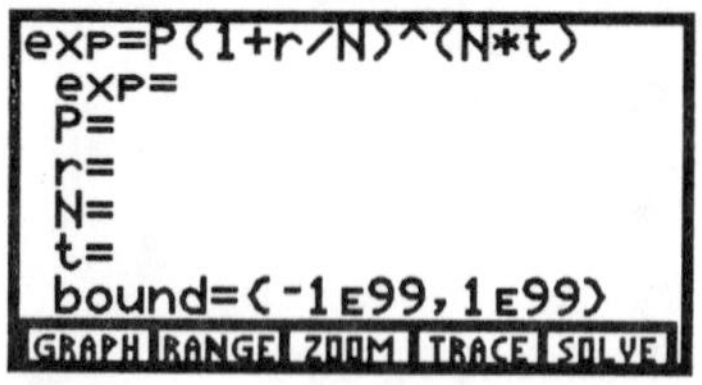

```
exp=P(1+r/N)^(N*t)
 exp=
 P=
 r=
 N=
 t=
 bound=(-1E99,1E99)
GRAPH|RANGE|ZOOM|TRACE|SOLVE
```
Clear any existing data beside the variables.

Enter the following data:

```
exp=P(1+r/N)^(N*t)
 exp=
 P=8000
 r=.08
 N=12
 t=2
 bound=(-1E99,1E99)
GRAPH|RANGE|ZOOM|TRACE|SOLVE
```

Now move the cursor beside "exp=" and then press **SOLVE** (F5).

```
exp=P(1+r/N)^(N*t)
•exp=9383.1034539704
 P=8000
 r=.08
 N=12
 t=2
 bound=(-1E99,1E99)
GRAPH|RANGE|ZOOM|TRACE|SOLVE
```

The future value of an $8000 investment at 8% for 2 years compounded monthly is $9383.10.

We may use the SOLVER to solve for any of the variables in the formula.

If the compounding is continuous, use the F2 formula to get:

CALCULATOR INSTRUCTIONS IV

Transmitting A Program From One Calculator To Another
and
Using The Table Program For The TI-85

I. Transmitting A Program (Connect both calculators with cable)

<u>Receiving Calculator</u>   <u>Transmitting Calculator</u>

LINK (2nd x-VAR)   LINK (2nd x-VAR)
RECV (F2)   SEND
You should see "Waiting"   PRGM (F2)
                                    Position pointer beside the program to
                                    be sent
                                    SELCT (F2)
                                    XMIT (F1)
                                    You should see "Done"

II. Using The Table Program For The TI-85 (Get From Your Instructor using directions in I) At this writing, the table program used is the one written by Prof. Mark Janeba, Dept. of Math, Willamette Univ, Salem, OR 97301.

Before using the TABLE program, the function(s) that you wish to use MUST be entered as y1 to y4 in the GRAPH editor.

For now, we will use the following functions:

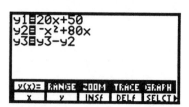

Once the function(s) to be used first is in the y1-y4 position, follow the following steps:

EXIT, EXIT
PRGM
NAMES (F1)
TABLE - Should be above one of the menu buttons.
ENTER

You should see the following:

> Your function(s)
> should be in y1..y4.
>
> Range | #Fns | #digt | Table | quit

Press F1 (Range)
Type in "0" beside TblMin = , press ENTER, type in "1" beside ΔTbl =, press ENTER. This process set the table beginning at x = 0 and increases x by increments of 1.

Press #Fns (F2)
Now answer the question "How many functions?" We are using 3. Type 3, ENTER. (The default is 1.)

Press #digt (F3)
We will use 2. Type 2, ENTER (The default is "-1" for floating.) If the numbers have many digits, the columns overlap.

Press Table (F4) to see the following:

```
x       y1      y2
0.00    50.00   0.00
1.00    70.00   79.00
2.00    90.00   156.00
3.00    110.00  231.00
4.00    130.00  304.00
5.00    150.00  375.00
PageD | PageU | NxtFn | PrvFn | Main
```

To see the values for y3, press NxtFn (F3).

```
x       y2      y3
0.00    0.00    50.00
1.00    79.00   -9.00
2.00    156.00  -66.00
3.00    231.00  -121.00
4.00    304.00  -174.00
5.00    375.00  -225.00
PageD | PageU | NxtFn | PrvFn | Main
```

The sub-menu for the Table menu is as follows:

- PageD: Shows the next 6 entries in the table (increasing x).
- PageU: Shows the previous 6 entries in the table (decreasing x)
- NxtFn: Shows the next column to the right using the next function. This wraps back to y1 when you run out of function.
- PrvFn: Opposite of NxtFn. Shows next column to the left.
- Main: Returns you to the main menu, where settings can be changed.

Notes:

1. The program crashes if you enter a larger number (1,2,3, or 4) than you have functions in the GRAPH editor.
2. Use "quit" (F5) to exit the program.

TI-82: CALCULATOR INSTRUCTIONS I

OBJECTIVES: The student should be able to enter and analyze statistical data.

This will include:
1. entering data points (one or more sets) in the STAT editor,
2. graphing, using STAT PLOT, to display a scatter plot(s) and an xyLINE graph(s),
3. performing regression analyses (curve fitting),
4. forecasting a statistical data value.

1. Entering data points in the STAT editor:

From the STAT menu select EDIT (press 1)

a. The top line displays three of the six possible list names L1, L2, ...L6.
b. The center portion of the screen will display up to seven elements of a list. If the element is to long to display, an abbreviated form will be shown. The full value of the element indicated in the rectangular cursor is shown on the bottom line. If your new list is shorter than the previously entered list under this list name, you can clear this list by pressing STAT, 4, then type in the list name and press ENTER. NOTE: the list names L1, L2, ... L6 are 2nd function keys above the numbers 1,2, ... 6.

Example 1:

Enter the following data points in the STAT editor.

x	-1	0	2	5	8
y	3	-1	0	2	5

Press STAT, EDIT (1)

a. Use L1 for the x element list and L2 for the y element list. You may need to ClrLIST before entering data.
b. Enter -1,0,2,5,8 pressing ENTER after each number.
c. Now press the right arrow key to move over to L2 for the y elements.
d. Enter 3,-1,0,2,5 pressing ENTER after each number.

APPENDIX A

Example 2:

Enter the following data points in the STAT editor.

x	-5	0	1	2	6
y	-1	1	3	2	-2

Press STAT, EDIT (1)

a. Use L3 for the x list and L4 for the y list.
b. Enter the numbers -5,0,1,2,6 pressing ENTER after each number under L3.
c. Enter the numbers -1,1,3,2,-2 pressing ENTER after each number under L4.

2. **Graphing, using STAT PLOT (2nd y=), to display a scatter plot(s) and an xyLINE graph(s):**

    NOTE: The STAT PLOT is closely connected to the GRAPH editor.
    1. You must clear or deselect any functions listed under the **y=** menu key or they will be plotted.
    2. You must define the viewing window.

To display a scatter plot or xyLINE graph of the data points in Example 1:

a. Press **y=** and clear or deselect any equations.
b. Press WINDOW. Since we will eventually draw Example 1 and Example 2 on the same window, we will now set the RANGE as follows:

    xMin = -6   yMin = -3
    xMax = 10   yMax = 8
    xScl = 1   yScl = 1

c. Press STAT PLOT (2nd Y=)
   Select Plot 1. Press ENTER.
d. Now use the arrow keys to move about in defining Plot 1. Select the following:
   ON
   Type: Scatter or xyLINE
   xlist: L1
   ylist: L2
   Mark: small open rectangle
e. Press GRAPH
   You should now see the following graph:

APPENDIX A

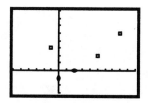

To display an xyLINE graph for Example 1: (Plots and connects data points with lines)

Press STAT PLOT, ENTER.
Change the type graph to xyLine, then press GRAPH.
Your graph should look like the following:

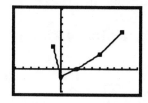

To display an xyLINE graph for both Example 1 and Example 2 on the same screen:

Press STAT PLOT. Move cursor to Plot 2, press ENTER.
Define Plot 2 as follows:
    ON
    Type: xyLine
    xList: L3
    yList: L4
    Mark: open rectangle
Press GRAPH
Your graph should be as follows:

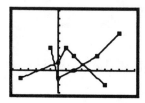

3. Performing regression analyses (curve fitting):

After selecting or entering the lists to use, select CALC from the STAT editor. Select SET UP to assign the names of the xLIST and yLIST. Then press STAT then CALC again. Enter the statistical calculation of choice by typing the number that appears beside the regression that you need. The analysis is calculated using the least-squares fit.

Key	Analysis	Regression Equation
1	1-Var Stats	
2	2-Var Stats	
3	Set UP...	
4	Med-Med	
5	LinReg(ax+b)	$y=ax+b$
6	QuadReg	$y=ax^2+bx+c$
7	CubicReg	$y=ax^3+bx^2+cx+d$
8	QuartReg	$y=ax^4+bx^3+cx^2+dx+e$
9	LinReg(a+bx)	$y=a+bx$
0	LnReg	$y = a + b \ln(x), \ x>0$
A	ExpReg	$y = ab^x, y > 0$
B	PwrReg	$y = ax^b, x > 0, y > 0$

For two-variable regression lists, except polynomial regression, **r** (correlation coefficient) measures the goodness of the fit of the equation with the data points. In most cases, the closer **r** is to 1 or -1, the better the fit. If **r** = 0 then x and y are completely independent.

Example 3:

A company's annual sales for the first five years in business is given by the following table.

x	1	2	3	4	5
y	7	10	11	14	15

Find the equation of the least squares line. What is the correlation coefficient?

Step 1: Enter data points in STAT, EDIT under two of the six list names.
Step 2: Press STAT,CALC, #3(SETUP), ENTER. Set up the xList and yList names.
Step 3: Press STAT,CALC, #9(Linear Regression), ENTER. Your screen should read as follows:

Your equation is y = 5.4 + 2x   r = .985329 (pretty good fit)

To display the regression curve in the STAT editor:

a. Press STAT PLOT (2nd y=), ENTER (I will use Plot 1). Make sure no other p l o t s are on.
b. Setup Plot 1 as follows:

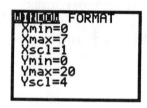

c. Press Y=. Type in the rest of the regression equation "5.4 + 2x".
d. Press WINDOW and setup.

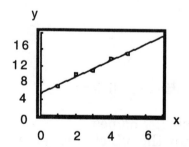

e. Press GRAPH.
   Your graph should be as follows:

Example 5:

Perform regression analyses for the data points in Example 3 using :

a. Logarithmic Regression
b. Exponential Regression
c. Power Regression

Compare the correlation coefficients. Which regression of these and the linear, seems to be the "best fit"? Now draw the "best fit" curve and the scatter plot. Your window should be similar to the following:

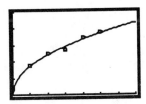

Power regression $y = ax^b$, $y = 7.0015x^{.47266}$ with corr. = .990697

4. Forecasting a statistical data value:

a. Press TblSet (2nd WINDOW)
   Setup the min. for x, the increments for x, auto,auto.
b. Press TABLE (2nd GRAPH) and move, using arrow keys, the cursor to the value of x that you wish to solve the regression curve for. Move the cursor to the right and read the Y_1 value at the bottom of the screen.

Example 6:

Using the power regression in Example 5, forecast the sales for x=6. Your screen should be as follows:

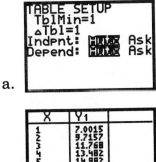

a.

b.

APPENDIX A   xxv   TI-82: I

TI-82: CALCULATOR INSTRUCTIONS II

OBJECTIVES: The student should be able to enter and graph functions.

This will include:

1. entering a function(s) in the Y= menu,
2. setting an appropriate viewing window,
3. displaying the graph,
4. exploring the graph with the TRACE feature,
5. exploring the graph with the ZOOM features,
6. analyzing the graph with the CALC features.

1. Entering a function(s) in the Y= menu:
a. Press the Y= key.
b. Type the function(s) beside the appropriate Y= , using the **X,T,Ø** key to represent the independent variable. **NOTE: THIS VARIABLE KEY WILL BE CALLED THE "X" KEY.**
c. Press ENTER after each function is typed in.

2. Setting an appropriate viewing window:
Press the WINDOW key and enter the appropriate numbers.

3. Displaying the graph:
Press the GRAPH key.

4. Exploring the graph with the TRACE feature:
Press the TRACE key. The blinking cursor shows the location of the point given by the x and y coordinates at the bottom of the screen. Use the left or right arrow keys to move around the screen on the curve. Panning left or right off the display will automatically change the viewing window and update the WINDOW settings. While using the trace feature, pressing the ENTER key will adjust the viewing window so that the cursor will be in the center of the screen. The WINDOW will automatically be updated.

5. **Exploring the graph with the ZOOM features:**
There are nine ZOOM features. You can explore all of these in your calculator guidebook. The following is a list of the features that we will use with only a brief description:

Feature	Brief Description
ZBox	Boxes in an area of the curve (cursor defined), then redraws the graph on a new screen - showing only the area in the box.
Zoom In	Magnifies the graph centered around the cursor location.
Zoom Out	Displays more of the graph.
ZStandard	Displays the graph using the default WINDOW.

6. **Analyzing the graph with the CALC features:**
There are seven CALC features. You can explore all of these in your calculator guidebook. The following is a list of the features with a brief description.

Feature	Description
value	Evaluates y when x= some given value.
root	Finds the roots of a function in a given interval.
minimum	Finds the minimum of a function on a given interval.
maximum	Finds the maximum of a function on a given interval.
intersect	Finds the intersection of two functions in a given interval.
dy/dx	Finds derivative (slope) of a function at a given point.
∫ f(x)	Finds the definite integral of a function in a given interval.

Example 1:

Graph   a. **y = (3/5)x - 4** and b. **y = -2x + 6** on the same screen.
Find the following for both a and b by using the calculator features:

1. y-intercept (x=0)
2. x-intercept (root)
3. intersection of a and b

Solution:

1. Press the Y= key. **NOTE:** We will use the key "XTØ" to type in a variable in an equation. When you see "X" or "x" the key "XTØ" should be used.
2. Beside "$Y_1 =$" type $(3/5)x - 4$ by pressing the following keys:
$(3 \div 5)X - 4$ then ENTER. Beside "$Y_2 =$" type the other equation.
3. Press WINDOW and use the default setting or press ZOOM then ZStandard (6).
4. When using the ZStandard, the graph will display immediately after pressing 6. If you do not use the ZStandard then press the GRAPH key.

Your screen should be as follows:

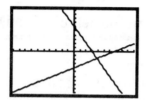

Now to use the CALC features to find:

1. the y-intercept for:

    graph a) - Press CALC (2nd TRACE), 1, and your screen should look as follows:

    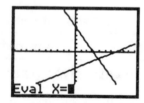

    NOTE: the y-intercept occurs when x=0. Type in 0 and then press ENTER. Your screen should look as follows:

    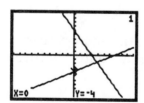 The y-intercept for graph a (function Y_1) is -4.

    For the y-intercept of graph b (function Y_2) press the up arrow button to get the following:

APPENDIX A xxviii TI-82: II

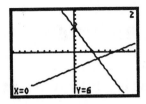 The y-intercept is 6.

2. the x-intercept for

graph a) - Press the CALC key (2nd TRACE), then 2 (root) and our screen should look as follows after moving the cursor to set a lower bound, an upper bound, and a guess.

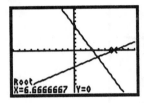 The x-intercept (root) for graph "a" (function Y_1)= 6.66666667.

graph b) - Press the CALC key (2nd TRACE), then 2 (root), upper arrow key (for function two), our screen should look as follows after moving the cursor to set a lower bound, an upper bound, and a guess. :

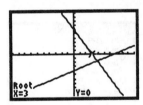 The x-intercept (root) for graph "b" (function Y_2)=3.

3) the intersection of the lines

Press the CALC key (2nd TRACE), then 5 (intersect). Move arrow key to what looks like the intersect point. Press ENTER 3 times. Our screen should look as follows:

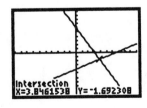 The intersection of Y_1 and Y_2 is (3.8461538, -1.692308)

TI-82 CALCULATOR INSTRUCTIONS II-B

These are Examples 3 and 4 in section 2.1. This is a little different approach.

Example 3: Graph y=-110x+15

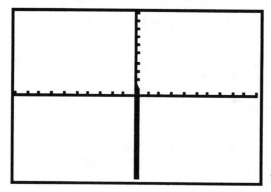 using ZStandard

Knowing that the y intercept is 15 and the x-intercept is 15/110 which is < 1, we

might set our RANGE window to the following: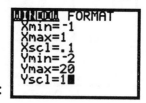

We would then get [graph].

APPENDIX A   xxx   TI-82: II-B

Example 4: Graph $y = \dfrac{7}{327} x + \dfrac{23}{327}$

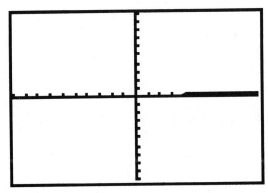 using ZStandard

Knowing that the y-intercept is very close to 0 (i. e. 23/327), manually change the Range Window to the following:

 to get the following graph 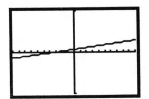.

Leaving the yScl = 1 instead of changing it to .1, merely removes the tick marks between -1 and 1 on the y-axis. The graph looks the same.

APPENDIX A   xxxi   TI-82: II-B

TI-82: CALCULATOR INSTRUCTIONS III

The TI-82 does not have the sophisticated solver that is available on the TI-85. Sections 4.1 through 4.3 will require more effort and work using the TI-82 calculator. The following is a suggestion for using the TI-82 with some of the examples in 4.1 and 4.3.

Section 4.1

Example 1: Finding future values

Find the future value of $1,000 at 6% interest for 20 years compounded:

- a. annually
- b. semiannually
- c. quarterly
- d. monthly
- e. daily
- f. continuously

For parts a through e we will use the F1 formula **F1 = P(1+(r/N))^(N*t)**. P = $1,000, r = .06, t=20, we will let Y1 represent F1, and x = N since N is the variable that is changing. Follow the following steps:

1. Press Y=
   Beside Y1(or some Y#)= type: 1000(1+(.06/X))^(20X)

2. Press WINDOW and set as follows:
   Xmin = 0
   Xmax = 12
   xscl = 1
   Ymin = 0
   Ymax = 5000
   Yscl = 500

3. Press GRAPH to see the following:

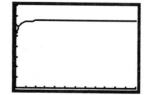

4. Press CALC (2nd TRACE)
   Select 1:value and press ENTER
   Enter "1" beside x= for annually, then press ENTER to see the following:

APPENDIX A

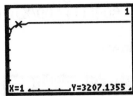

The future value when the compounding is done on an annual basis is $3207.14.

Repeat step 4, letting x = 2 for semiannually, x = 4 for quarterly, and x = 12 for monthly. To find the future value for daily compounding we must change the WINDOW to include x = 365. The following shows these future values:

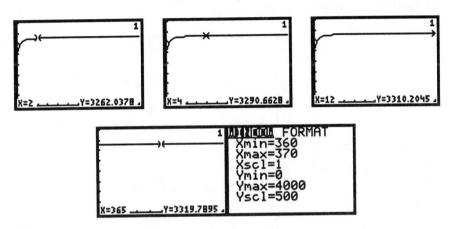

For continuous compounding we will use the F2 formula **F2 = Pe^(rt)**.

On the HOME screen:

Example 2: Finding present values

Find the present value of $6,000 at 9% interest for 10 years compounded:

  a. annually
  b. semiannually
  c. quarterly
  d. monthly
  e. daily
  f. continuously

For parts a through e we will use the present value formula **P = F1(1+(r/N))^(-N*t)**. F1 = $6,000, r = .09, t = 10, we will let Y2 represent P, and X = N since N is the variable that is changing. Follow the following steps:

1. Press Y=

APPENDIX A　　　　　　　　　　xxxiii　　　　　　　　　　TI-82: III

Beside Y2 (or some Y#) = type: $6000(1+(.09/X))^{\wedge}((\text{-})10X)$

2. Press WINDOW and set as follows:

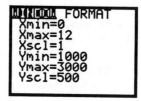

3. Press GRAPH to see the following:

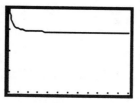

4. Press CALC (2nd TRACE)
   Select 1:value and press ENTER
   Enter "1" beside x= for annually, then press ENTER to see the following:

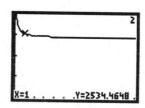

The present value when the compounding is done on an annual basis is $2534.46.

Repeat step 4, letting x = 2 for semiannually, x = 4 for quarterly, and x = 12 for monthly. To find the present value for daily compounding we must change the WINDOW to include x = 365. The following shows these present values:

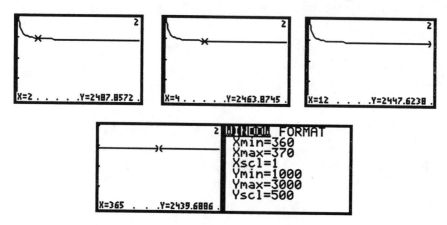

APPENDIX A   xxxiv   TI-82: III

For continuous compounding we will use the present value formula **P = F2(e^(-r*t))**.

On the HOME screen:

Section 4.3

Example 2:

Find the monthly payment and the total interest paid, for a loan of $5000 at a rate of 11.75% per annum, if the loan is to be paid back in

a. 3 years
b. 4 years

Solution: Use the LP formula letting L = 5000, r = .1175, N = 12, and t = X.
a.
1. Press Y=
   Type in the following beside a Y# (I used Y4)

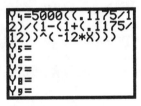

2. WINDOW setting:

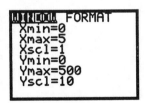

3. GRAPH

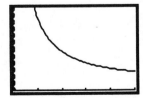

4. CALC
   1: value, ENTER
   let x = 3, ENTER

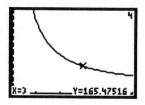

For a 3 year loan the monthly payments are $165.48.

Total interest paid = [$165.48*(12*3)] − $5000 = $957.28.

b.
   CALC
   1: value, ENTER
   let x = 4, ENTER

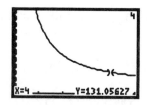

For a 4 year loan the monthly payments are $131.06.

Total interest paid = [$131.06*(12*4)] − $5000 = $1290.

The graph of the monthly payment (Y4) vs. t in years (X) and the total interest paid, letting LP be Y4 and t be X, is as follows:

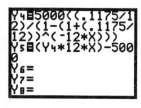 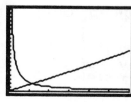 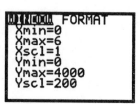

NOTE: To type Y4 in the Y5 equation press Y-VARS (2nd VARS), 1: Function, ENTER, 4.

Using a TABLE, we get the following:

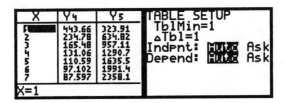

We can see from this that if the loan is to be paid back in monthly payments for 6 years, the monthly payment will be $97.10 and the total interest paid over the 6 years is $1991.35.

ANSWERS AND SOLUTIONS

Exercises 1.1

1.

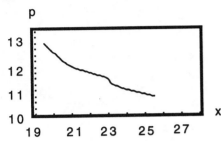

or

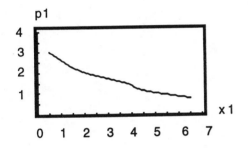

x1=x-19
p1=p-10

3.

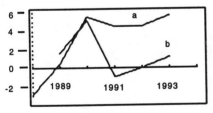

a. y1=y+29
b. y2=y

Exercises 1.2

1. a. quadratic   b. cubic
3. a. exponential or cubic   b. cubic, no

Exercises 1.3

1. a. $y = .3071x^2 - 2.9100x + 2000.4457$
   b. $y = -.1500x^3 + 3.4571x^2 - 24.4500x + 2048.3257$
3. $y = -.3000x^3 + 6.264x^2 - 42.25x + 107.1914$
5. $y = 55 - .00166x$, 13.5
7. $y = .2139x^3 - 2.0726x^2 + 3.9992x + 6$

Exercises 2.1

1. a. $m = \dfrac{14,000}{-150} = \dfrac{280}{-3}$ b. $y - 17,000 = \dfrac{-280}{3}(x - 400)$ c. $y = \dfrac{-280}{3}x + \dfrac{163,000}{3}$

3. $y = 54333.3333 - 93.333x$, yes

5. $p - 5 = -.001(q - 4000)$ therefore $p = -.001q + 9$
   substitute 4 for p and solve for q
   $q = 5000$

Exercises 2.2

1. Answer in the problem
3. M.E. = (225, $324.69)
5. S: $p = 5,500,000 + 60,000q$ D: $p = 14,000,000 - 100000q$
   M.E. = (53, $8,687500)

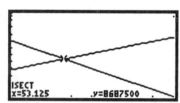

7. M.E. = (168, $.73)

9. D_o: $p = -.08875q + 73.2$ passes through (560, 23.50) and (600, 19.95)
   S_n: $p = .1014285714q - 40.9071428571$ passes through (635, 23.50) and (600, 19.95)
   S_o: $p = .1014285714q - 33.3$ has same slope as S_n and passes through (560, 23.50)
   D_n: Same as D_o

11. Too few, 20 demanded, 12 supplied.

13.

q	100	150	200	250	300	500
demand p	3.00	2.50	2.40	**2.19**	2.00	**1.03**
supply p	1.10	1.50	1.70	**2.03**	2.33	3.53

M.E. = (264.34, 2.119)

Exercises 2.3

1. a. P(x)=24x-(8x+2600)=16x - 2600,   b. P(500)=$5400,   c. P(x)=0 when x=162.5

   d. 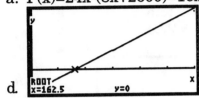   e.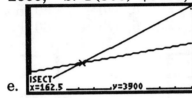

3. a. slope=4.5, intercept= -548, b. slope is additional profit per item sold, intercept is constant reduction of profit due to fixed cost.

5. C(x)=3.5x+32000, R(x)=5.95x, P(x)=R(x)-C(x)=5.95x - (3.5x + 32000)=2.45x-32000, solve P(x)=16000 to get x=19591.84

7. a. C(248)=$6802,   b. $354;   c. $26

9. a. C(x)=9.51x+380;   b. P(x)=30x-(9.51x+380)=20.49x-380, 0≤x≤400;

   c. x=18.545   d.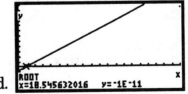

11. a. R(x) = .25x   b. P(x) = .25x − C(x)

13. a. P(x) = 13x - 3200   b. 700   c. x = 246.15

Exercises 2.4

1. a. $P(x) = (-19/18)x^2 + 850x - 31{,}000$ b. $x = 38.291398191774$ or $x = 766.97175970293$

3. $x = 1.357890677326$ and $p = 15.119296892442$

5. $q = 160.6042167782$ $f(q) = g(q) = 6.1007027963677$ at this q

7. $r = .080805$ $y = 4.9530324244$

9. D: $p = .002499999999931x^2 - .4403846153812x + 34.288461538407$
   S: $p = 3.79662070056282x^{.48293834680879}$

   a. D: when $q = 20$ then $p = 26.5$, when $q = 50$ then $p = 18.5$
      S: when $q = 40$ then $p = 22.5$, when $q = 60$ then $p = 27.4$

   b.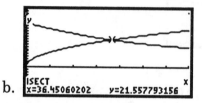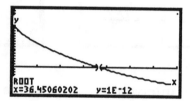

   c. $R(q) = .002499999999931x^3 - .440384615381x^2 + 34.288461538407x$

Exercises 2.5

1. a. $p = 1.97231930010326 - .00055043685464654x$

   b.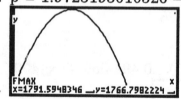

3. a. $C(q) = .8q + 150$ and $R(q) = .9x$ therefore break-even $= 1500$
   b. at $.90 the quantity demanded is 1948 which yields a profit for the supplier of $44.80

5. 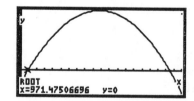   and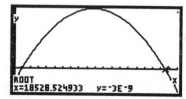

APPENDIX B   4   Ans. 2.5

7. a. $P(q) = 10q - .1q^2 - (7+5q+.1q^2) = 5q - .2q^2 - 7$

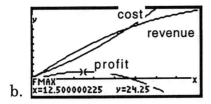
b.

c. $q = 12.5$ $P = 24.25$

d. revenue

9. a. $R(q) = 177q - .000666666667q^2$
   b. $q = 289,286$ $R(q) = \$12,150,000$

Exercise 3.1

1. a. $C(q) = .04q^2 - 12q + 1369$, $R(q) = 4.32q$, $P(q) = -.04q^2 + 16.32q - 1369$

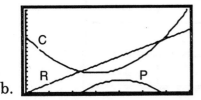
b.
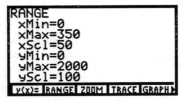

c. Use GRAPH, MORE, MORE, F1: Put in the x, then use the up and down arrow keys to evaluate the y for the three different functions.

q	129	130	179	180	229	230	279	280
cost	486.64	485	502.64	505	718.64	725	1134.64	1145
rev	557.28	561.6	773.28	777.6	989.28	993.6	1205.28	1209.6
profit	70.64	76.6	270.64	272.6	270.64	268.6	70.64	64.6
MP		5.96		1.96		-2.04		-6.04

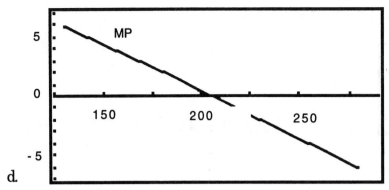

d. MP = 0 when q = 204, yes

3.

units	400	401	402	403	404	405
cost	631.35	631.74	632.14	632.54	632.95	633.85
revenue	800	802	804	806	808	810
profit	168.65	170.26	171.86	173.46	175.05	176.15
marginal profit		1.61	1.6	1.6	1.59	1.1

Increase production

Exercises 3.2

1. a. Plant I

q	70	71	72
C	285	276	269
R	297.5	301.75	306
P	12.5	25.75	37
MP		13.25	11.25

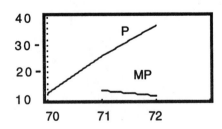

increase q up from 72

b. Plant II

q	79	80	81
C	276	285	296
R	335.75	340	344.25
P	59.75	55	48.25
MP		-4.75	-6.75
P-45	14.75	10	3.25

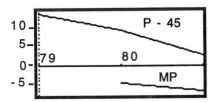

decrease q down from 80

3.

q	129	130	179	180	229	230	279	280
P	70.64	76.6	270.64	272.6	270.64	268.6	70.64	64.6
MP		5.96		1.96		-2.04		-6.04
MP*20		119.2		39.2		-40.8		-120.8

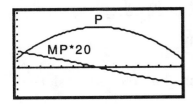 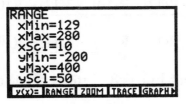

APPENDIX B   Ans. 3.2

5. a-b

q	400	401	402	403	404	405	406	407	408
C	631.35	632.35	633.5	634.75	636.1	637.5	639	640.4	641.7
R	800	802	804	806	808	810	812	814	816
P	168.65	169.65	170.5	171.25	171.9	172.5	173	173.6	174.3
MP		1	0.85	0.75	0.65	0.60	0.5	0.60	0.70
MP * 14		14	11.90	10.5	9.10	8.40	7	8.40	9.80
P - 150	18.65	19.65	20.5	21.25	21.9	22.5	23	23.6	24.3

c.

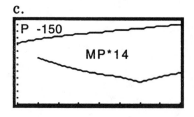 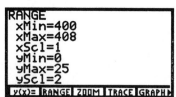

d. increase production up from 408

Exercises 3.3

1. $R(q) = 3q$

$$\frac{R(13) - R(10)}{13 - 10} = \frac{R(13) - R(10)}{3} = \frac{3(13) - 3(10)}{3} = \frac{39 - 30}{3} = 3$$

$$\frac{R(12) - R(10)}{12 - 10} = \frac{R(12) - R(10)}{2} = \frac{3(12) - 3(10)}{2} = \frac{36 - 30}{2} = 3$$

$$\frac{R(11) - R(10)}{11 - 10} = \frac{R(11) - R(10)}{1} = \frac{3(11) - 3(10)}{1} = 3$$

at q = 10 the slope of the tangent line is 3. Linear functions have constant slopes.

3. a. t = 1.5 yrs., 12

   b. $t \varepsilon$ (0, 1.5) marginal sales are growing

   $t \varepsilon$ (1.5, 3) marginal slaes are falling

Exercises 3.4

1. $f'(x) = \lim\limits_{h \to 0} \dfrac{[5(x+h)+8]-[5x+8]}{h} = \lim\limits_{h \to 0} \dfrac{5x+5h+8-5x-8}{h} = \lim\limits_{h \to 0} \dfrac{5h}{h} = \lim\limits_{h \to 0} 5 = 5$

3. $f'(x) = \lim\limits_{h \to 0} \dfrac{[2(x+h)^2+3]-[2x^2+3]}{h} = \lim\limits_{h \to 0} \dfrac{2x^2+4xh+2h^2+3-2x^2-3}{h} =$

   $\lim\limits_{h \to 0} \dfrac{4xh+2h^2}{h} = \lim\limits_{h \to 0} 4x+2h = 4x$

5. $\lim\limits_{h \to 0} \dfrac{\pi - \pi}{h} = \lim\limits_{h \to 0} \dfrac{0}{h} = \lim\limits_{h \to 0} 0 = 0$

7. $f'(x) = \lim\limits_{h \to 0} \dfrac{[(x+h)+1]^2-(x+1)^2}{h} = \lim\limits_{h \to 0} \dfrac{(x^2+2xh+2x+2h+h^2+1)-(x^2+2x+1)}{h} =$

   $\lim\limits_{h \to 0} \dfrac{2xh+2h+h^2}{h} = \lim\limits_{h \to 0} 2x+2+h = 2x+2$

Exercises 3.5

1. $3x^2 + 8x$

3. $8x + 4 - 4x^{-2} - 8x^{-3}$

5. $18x + 6$

7. $192x^2 + 96x + 12$

9. $\frac{1}{2}x^{\frac{-1}{2}} + \frac{1}{5}x^{\frac{-4}{5}}$

Exercises 3.6

1. $(3x+4)(3) + (3x+4)(3)$

3. $\dfrac{(1+2x)(20x^3) - (5x^4)(2)}{(1+2x)^2}$

5. $\sqrt[3]{3}\left(\dfrac{1}{3}x^{\frac{-2}{3}}\right)$

7. $\dfrac{(4^{.03}q^{.03})(5q^4 - 1) - (q^5 - q + 1)(4^{.03}(.03q^{-.97}))}{(4q)^{.03*2}}$

Exercises 3.7

1. $50(x^2+1)^{49}(2x)$

3. $\dfrac{1}{3}(5x^2-2x+1)^{-\frac{2}{3}}(10x-2)$

5. $[(2x+1)^5][7(3x-2)^6(3)]+[(3x-2)^7][5(2x+1)^4(2)]$

7. $\dfrac{(2x)[7(5x+2)^6(5)]-[(5x+2)^7](2)}{(2x)^2}$

9. $[\sqrt{5x-4}\,][10(x^2-3x)^9(2x-3)]+[(x^2-3x)^{10}][\tfrac{1}{2}(5x-4)^{-\frac{1}{2}}(5)]$

Exercises 3.8

1.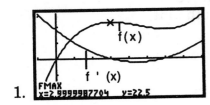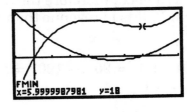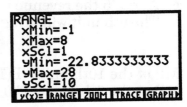

3.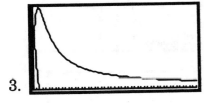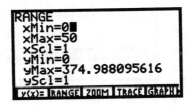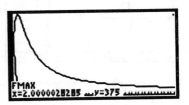

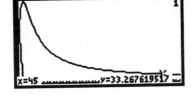

a. 3:17 PM   b.

c. 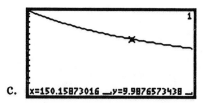 at 5:45 PM

5.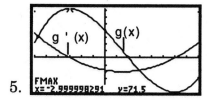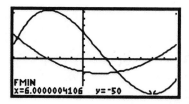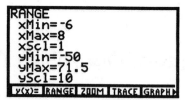

7. 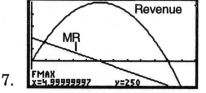 a. Max q=5 C(5)=250

   b. MR of the 6th item =R(6) - R(5) = 240 - 250 = -10
      MR at the 5th unit = R'(5) = 0

   c. At q = 5 the revenue is at a max., therefore the derivative is 0.
      The 6th unit will bring in 10 less in revenue than the 5th unit.

9. MP of the 10th unit = P(10) - P(9) = 2050 - 1907 = 143
   MP at the 9th unit = P'(9) = 151

11.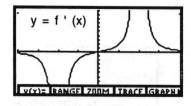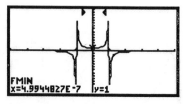

13 a. 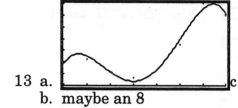 c.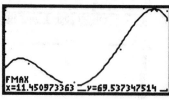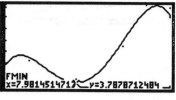
   b. maybe an 8

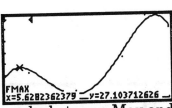

d. peaks between May and June, Nov. and Dec.

15.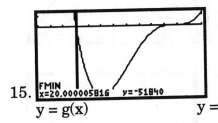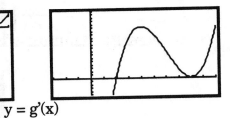
    $y = g(x)$ $y = g'(x)$

17.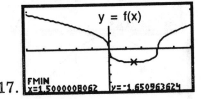

19.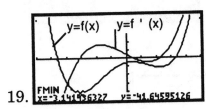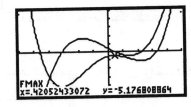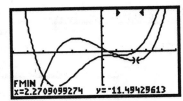

21.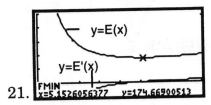

23. a. let x = #people over 200, but 200+x not to exceed 300
    $R(x) = (450 - 1.3x)(200 + x)$
    $C(x) = (200 + x)(200)$
    $P(x) = 90000 + 190x - 1.3x^2 - (40000 + 200x) = 50000 - 10x - 1.3x^2$

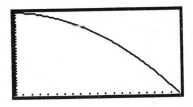

b. max. occurs at x=0, therefore 200 people   c. 450

APPENDIX B   13   Ans. 3.8

Exercises 3.9

1. a. $R(q) = (-4q+200)*q$ b. $C(q) = \sqrt{\dfrac{2000}{q}} * q + 500$

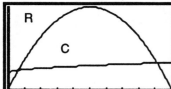
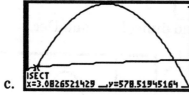
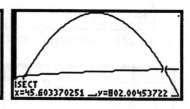

c.

d. $P(q) = R(q) - C(q) = -4q^2 + 200q - \left(\sqrt{\dfrac{2000}{q}} * q + 500\right)$ e. & f.

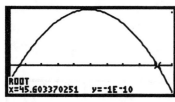
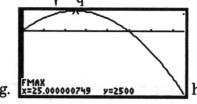
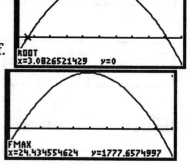

g. h.

i. revenue

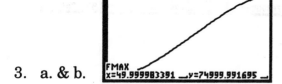

3. a. & b.

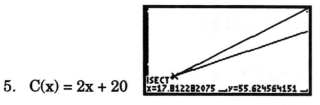

5. $C(x) = 2x + 20$

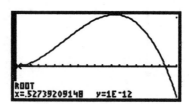

7. a. & b. $P(x) = (150x - 8x^2)x - (20x + 30)$

APPENDIX B   14   Ans. 3.9

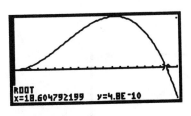

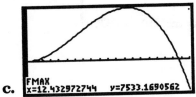
c.

d. P'(10)=$580 e. P(11)-P(10)=$482

9. a.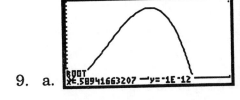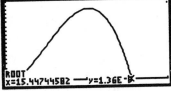

b.  c. P'(5)=$1260. d. P(6)-P(5)=$1254

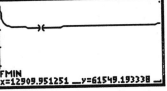

11. a. $C(x) = 50000*1.2+50000/x*200+.12*x/2$
    b. 5,000, C(5000)=$62,300, no.

13. a. $C(x) = 3.75*4000+\dfrac{4000}{x}*500+\dfrac{x}{2}*2.5$ b.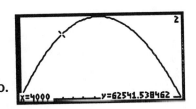

c. 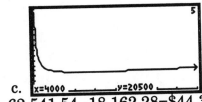 d.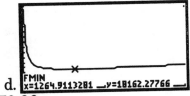

e. 62,541.54 -18,162.28=$44,379.26

APPENDIX B Ans. 3.9

15. $C(x) = 4.5*4000 + \dfrac{4000}{x}(500) + \dfrac{x}{2}(2.5)$ b.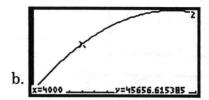

c.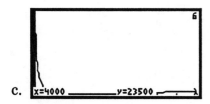

d. 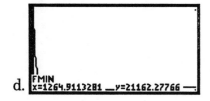   e. 45,656.62 − 21,162.28 = $24,494.34

Exercises 4.1

1.
```
exp=P*(1+(r/N))^(N*t)
 exp=2000000
■P=10836.300910222
 r=.11
 N=1
 t=50
 bound=(-1E99,1E99)
GRAPH RANGE ZOOM TRACE SOLVE
```

3. a.
```
exp=P*(1+(r/N))^(N*t)
■exp=4660.9571438495
 P=1000
 r=.08
 N=1
 t=20
 bound=(-1E99,1E99)
GRAPH RANGE ZOOM TRACE SOLVE
```

b.
```
exp=P*(1+(r/N))^(N*t)
■exp=4875.4391560972
 P=1000
 r=.08
 N=4
 t=20
 bound=(-1E99,1E99)
GRAPH RANGE ZOOM TRACE SOLVE
```

c.
```
exp=P*e^(r*t)
■exp=4953.0324243951
 P=1000
 r=.08
 t=20
 bound=(-1E99,1E99)
■left-rt=0
GRAPH RANGE ZOOM TRACE SOLVE
```

5.
```
exp=P*e^(r*t)
 exp=100000
■P=22313.016014843
 r=.075
 t=20
 bound=(-1E99,1E99)
■left-rt=0
GRAPH RANGE ZOOM TRACE SOLVE
```

7.
```
exp=P(1+r/N)^(N*t)
■exp=168897253.95307
 P=750
 r=.045
 N=1
 t=280
 bound=(-1E99,1E99)
GRAPH RANGE ZOOM TRACE SOLVE
```
for the first 280 years

```
exp=P(1+r/N)^(N*t)
■exp=348102571.16022
 P=168897253.95307
 r=.075
 N=1
 t=10
 bound=(-1E99,1E99)
GRAPH RANGE ZOOM TRACE SOLVE
```
at the end of the last 10 years

APPENDIX B　　　　　Ans. 4.1

Exercises 4.2

1. a.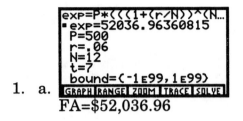
FA=$52,036.96   b. PA=$34,226.52

or use the answer in part "a" in F1 formula to solve for "P"

3. a.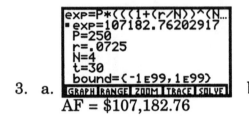
AF = $107,182.76   b.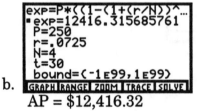
AP = $12,416.32

or use the answer in part "a" in F1 formula to solve for P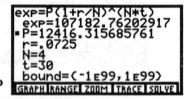

5. 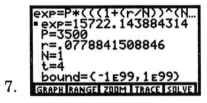   PA = $16,936.06

7. FA = $15,722.14

APPENDIX B   18   Ans. 4.2

9. a. John - $34,000   Paul - $87,500   b.

c. John has $74,747.67 to leave on deposit from age 41 to age 65. He ends up with the following:

John ends up with $542,368.81

Paul ends up with $428,292.01

11.

Exercises 4.3

1. Enter as y1 in the GRAPH editor the loan payment equation with L = 12000, r = .105, N = 12, and t = x-VAR.
   Enter as y2 the formula to calculate the total interest paid:
   y2 = (y1*12*x) - 12000
   Use the UT table program to generate the following:

```
 x      y1      y2              x       y1      y2
 6.00   225.35  4225.03          12.00   146.90  9153.15
 7.00   202.33  4995.56          13.00   141.30  10042.8
 8.00   185.28  5786.90          14.00   136.61  10950.8
 9.00   172.21  6598.72          15.00   132.65  11876.6
 10.00  161.92  7430.64          16.00   129.27  12819.6
 11.00  153.65  8282.26          17.00   126.37  13779.4
```

therefore reading from the table

a. if the loan is to be paid in 6 years, the monthly payment is $225.35 and the total interest paid is $4225.03.

b - g follows the same

3.
```
exp=L*((r/N)/(1-(1+(...
 exp=1070.4773034274
L=125000
r=.0925
N=12
t=25
bound=(-1E99,1E99)
```
For the 25 year loan the payment per month is $1070.48

```
exp=L*((r/N)/(1-(1+(...
 exp=1028.3442818812
L=125000
r=.0925
N=12
t=30
bound=(-1E99,1E99)
```
For the 30 year loan the payment per month is $1028.34.

```
1070.48*(12*25)
               321144
1028.34*(12*30)
               370202.4
370202.4-321144
               49058.4
```
The difference is $49,058.40

5.
```
exp=L*((r/N)/(1-(1+(...
 exp=845.8
L=110000
r=.085
N=12
 t=30.000788661602
bound=(1,1E99)
```
Note the bound Term is 30 years

7.
```
exp=L*((r/N)/(1-(1+(...
 exp=2385.5405876132
L=20000
r=.12
N=2
t=6
bound=(-1E99,1E99)
```
GRAPH editor y2 = 2385.5405876132 (semi-annual payment)

Set y3 = .12/2 = .06 (period rate)

Set y1 = y2((1−(1+y3)^(−(12−x)))/y3 (x = the payment #) We treat this as a present value of an annuity.

Use the UT table program to generate the following:

x	y1
0.00	20000.00
1.00	18814.46
2.00	17557.79
3.00	16225.71
4.00	14813.72
5.00	13317.00

x	y1
6.00	11730.48
7.00	10048.76
8.00	8266.15
9.00	6376.58
10.00	4373.63
11.00	2250.51

x	y1
12.00	0.00
13.00	−2385.54
14.00	−4914.21
15.00	−7594.61
16.00	−10435.82
17.00	−13447.51

Exercises 4.4

1. $3e^{3x}$

3. $3e^{3x+1}$

5. $6xe^{2x} + 3e^{2x}$

7. $e^{4x}\left(\dfrac{1}{x}\right) + \ln(4x) * 4e^{4x}$

9. $\dfrac{(x^3 + 1)[5xe^{x^2}(2x) + e^{x^2}(5)] - 5xe^{x^2}(3x^2)}{(x^3 + 1)^2}$

11. a.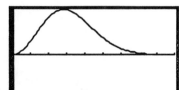

b. $R'(t) = \dfrac{100}{3!}\left[\left(t^2 e^{\frac{-t^2}{2}} * -t\right) + e^{\frac{-t^2}{2}}(2t)\right] = \dfrac{100}{3!} e^{\frac{-t^2}{2}}(2t - t^3)$

c.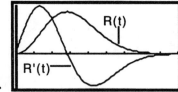

d. R'(t) is positive for $t \in (0, 1.4142135624)$ therefore R(t) is increasing. R'(t) is negative for $t \in (1.4142135624, \infty)$ therefore R(t) is decreasing.
R'(t) = 0 when t = 0 and 1.4142135624

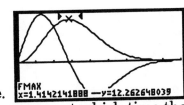

e. in a little over 1.4 years the max. sales rate will occur, at which time the sales will be 12,263 books a year.

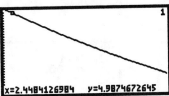

f. zooming in and tracing we get: in a little over 2yrs. and 5months, and

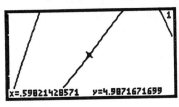

in a little over 7 months.

13. a. take the derivative and set equal to 0, then solve for x. x=c

   b. x=3

Graphs

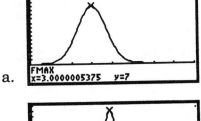

a.

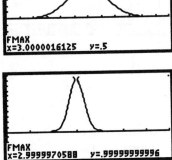

b.

c.

d.

APPENDIX B   23   Ans. 4.4

Exercises 4.5

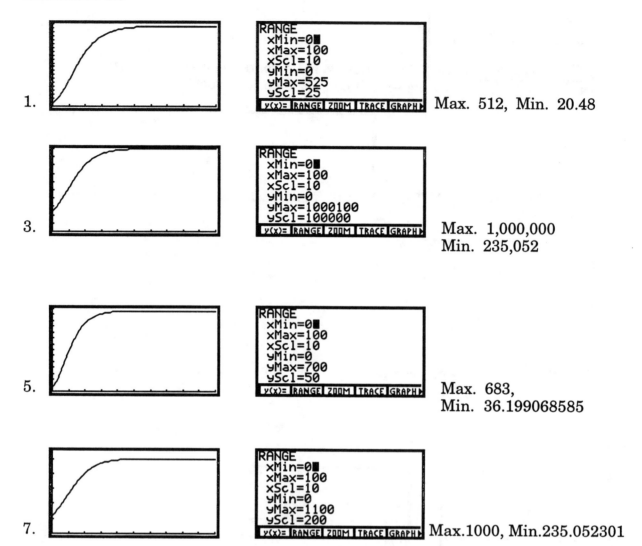

1. Max. 512, Min. 20.48

3. Max. 1,000,000
   Min. 235,052

5. Max. 683,
   Min. 36.199068585

7. Max. 1000, Min. 235.052301

EXERCISES 5.1

1. Discontinuous x = -4, missing point
3. Discontinuous x = 6, vertical asymptote
5. None
7. None

EXERCISES 5.2

1. $N(x) = \begin{cases} x - (500 + .23x) & 12{,}000 \leq x \leq 25{,}000 \\ x - (.5x - 6250) & 25{,}000 < x < \infty \end{cases}$

   $N(25{,}000) = 18750$ $N(25{,}001) = 18750.50$ Make more money, your income goes up.

3. No, it removes incentive at jumps.

5. a. $A(t) = \begin{cases} 8.5t & t \leq 8 \\ 68 + 12.75(t-8) & 8 < t \leq 16 \end{cases}$

   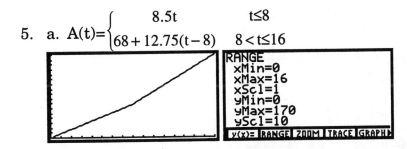

   b. $A'(t) = \begin{cases} 8.5 & 0 < x \leq 8 \\ 12.75 & 8 < x \leq 16 \end{cases}$

   c. Yes

   d. No

7. yes, no, no, no

9. a. $f(t) = \begin{cases} 3000e^{.05t} & t \leq 1 \\ 3000e^{.05(1)} * e^{.08(t-1)} & 1 \leq t \leq 3 \\ 3000e^{.05(1)} * e^{.08(2)} * e^{.11(t-3)} & 3 < t \leq 4 \\ 3000e^{.05(1)} * e^{.08(2)} * e^{.11(1)} * e^{.14(t-4)} & 4 < t \leq 5 \end{cases}$

   b.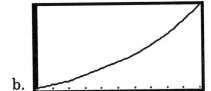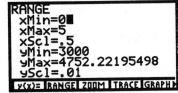

APPENDIX B   25   Ans. 5.2

11. a. $$C(t) = \begin{cases} .35 & t \leq 1 \\ .35 + (.12(\text{int}(t - .001))) & t > 1 \end{cases}$$ which is the same as

$C(t) = .35 + (.12(\text{int}(t - .001)))$ for $t > 0$

b. No

c. $C(.5) = .35 \quad C(1) = .35 \quad C(1.01) = .47 \quad C(1.99) = .47 \quad C(2) = .47$
$C(2.01) = .59$
$C(3.01) = .71$

d.

13. a. $$C(x) = \begin{cases} 50{,}000 * .85 + \left(\dfrac{50{,}000}{x}\right) * 40 + \left(\dfrac{x}{2}\right) * .05 & x \leq 10{,}000 \\ 50{,}000 * .765 + \left(\dfrac{50{,}000}{x}\right) * 150 + \left(\dfrac{x}{2}\right) * .05 & x > 10{,}000 \end{cases}$$

b & c.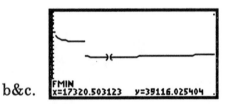

15. $C(t) = 17.32 + .10(\text{int}(t + .99))$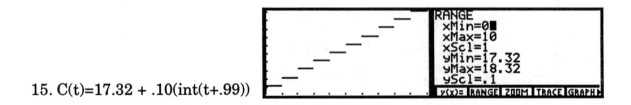

17. a. $C = 50{,}000(1.20) + (50{,}000/x)(200) + 0$ if $x \leq 2000$ $x = \#$ in each order

APPENDIX B
Ans. 5.2

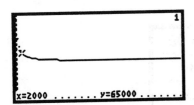

b. C = 50,000(1.20) + (50,000/x)(200) + .12(x/2) if 2000<x≤50000

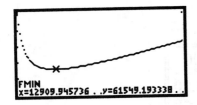

c. answer in b

19. a. $C(x) = \begin{cases} 50,000 + \dfrac{50000}{x}(200) & \text{if } x < 5000 \\ 50,000(.75) + \dfrac{50000}{x}(200) + .10\left(\dfrac{x}{2}\right) & \text{if } x \geq 5000 \end{cases}$

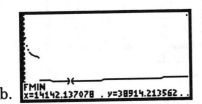

b.

c. 1. 4   2. three orders of 14,000 and one of 8,000   3. about 39,000

Exercises 5.3

1. a. E=.6933   E<1 therefore inelastic
   b. increase price to increase revenue; but not so much as to make demand elastic or to lengthen selling time to much

3. a. E=1   b. leave the price at .65/dip

5.

p	18.99	19.99	23.99	25.99	29.99
q	700	680	550	470	80
E		.56	-2.01	1.96	9.92
R	13,293	13,593.20	13,194.50	12,215.30	2,399.20

a. p should be closer to 19.99

C	10,500	10,200	8,250	7,050	1,200
P	2,793	3,393.20	4,944.50	5,165.30	1,199.20

b. a max profit will occur around a price of 25.99

7.

Month	Oct.	Nov.	Dec.	Jan.	Feb.	Mar.
$39.00	*50	*275	1550	23	*51	*55
$29.50	55	350	*2200	*35	60	72
E	.33	.83	1.20	1.43	.56	.92

Note: a * is placed in the line of the best price for each month

```
RANGE
 xMin=1
 xMax=7
 xScl=1
 yMin=.3
 yMax=1.5
 yScl=.1
y(x)= RANGE ZOOM TRACE GRAPH
```
Elasticity on y-axis

Exercises 5.4

1. (1-2)*(-10)=-1*(-10)=10   therefore revenue is increased by 10%

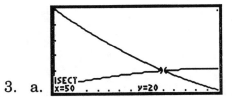

3. a.

   b. Using TRACE on the demand we get that q=46 wheb p=22

   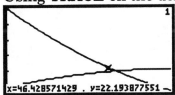

   finding the derivative of the supply when q=46, we get

   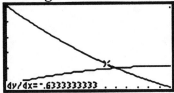

   therefore $\eta = -\dfrac{1}{-.63333333334}\left(\dfrac{22}{46}\right) = .75514871418$

   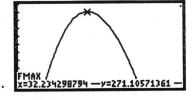

   c.

Exercises 5.5

1.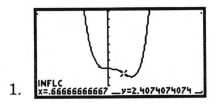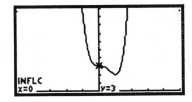

3.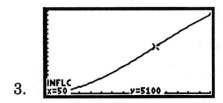

Exercises 6.1

1. $\dfrac{x^3}{3}+k$

3. $\dfrac{x^{-2+1}}{-2+1}=-x^{-1}+k$

5. $\dfrac{x^{\frac{3}{5}+1}}{\frac{3}{5}+1}=\dfrac{x^{\frac{8}{5}}}{\frac{8}{5}}=\dfrac{5}{8}x^{\frac{8}{5}}+k$

7. $\dfrac{x^{\frac{-3}{5}+1}}{\frac{-3}{5}+1}=\dfrac{x^{\frac{2}{5}}}{\frac{2}{5}}=\dfrac{5}{2}x^{\frac{2}{5}}+k$

9. $\dfrac{4x^{-3+1}}{-3+1}=-2x^{-2}+k$

11. $6x+k$

13. $\dfrac{x^2}{2}+3x+k$

15. $x^4-2x^3+\dfrac{5}{2}x^2-10x+k$

17. $\dfrac{3x^{\frac{1}{2}+1}}{\frac{1}{2}+1}-\dfrac{3x^{-2+1}}{-2+1}=\dfrac{3x^{\frac{3}{2}}}{\frac{3}{2}}-\dfrac{3x^{-1}}{-1}$

    $2x^{\frac{3}{2}}+3x^{-1}+k$

19. $-3e^x+k$

21. $12\ln|x|+k$

23.

$$C = \int (2x^3 + 5x - 6)dx = \frac{2x^4}{4} + \frac{5x^2}{2} - 6x + k = \frac{1}{2}x^4 + \frac{5}{2}x^2 - 6x + k$$

FC = \$750 i.e. C(0) = 750 $\therefore 750 = \frac{1}{2}0^4 + \frac{5}{2}0^2 - 60 + k$ $\therefore 750 = k$

$$C(x) = \frac{1}{2}x^4 + \frac{5}{2}x^2 - 6x + 750$$

25.

$$C = \int (x^2 - 2x + 3) dx = \frac{x^3}{3} - x^2 + 3x + k, \quad C(3) = 15$$

$$15 = \frac{3^3}{3} - 3^2 + 3(3) + k, \quad k = 6$$

$$C(x) = \frac{x^3}{3} - x^2 + 3x + 6$$

27.

$$R = \int (100 - 6x - 2x^2) dx = 100x - 3x^2 - \frac{2}{3}x^3 + k$$

$R(0) = 0$ $\therefore k = 0$ $\therefore R(x) = 100x - 3x^2 - \frac{2}{3}x^3 =$

$(100 - 3x - \frac{2}{3}x^2)x$

Since R = Demand price * x

$$D(x) = 100 - 3x - \frac{2}{3}x^2$$

Exercise 6.2

1. a. 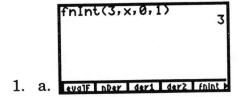 b.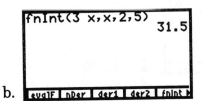

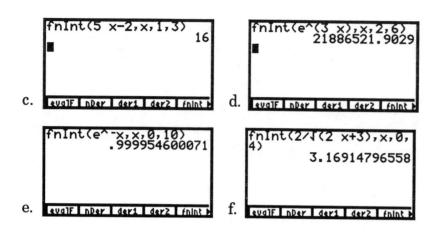
c.
d.
e.
f.

3. Store the equation in the graph editor as y1. Calculate the definite integral in the CALC editor

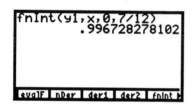

5.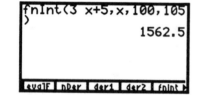

7. a. 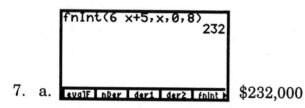 $232,000

b. 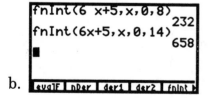 $658,000

Exercises 6.3

1. a. [calculator screens showing: $(1/\sqrt{(2\pi)})*\text{fnInt}(e^{\wedge}(-.5(x-7.5)^2),x,7.25,7.75)$ = .197412651366; Ans*.383 = .075609045473; Ans*50 = 3.78045227366] order 3 or 4 pairs

   b. [calculator screens showing: $(1/\sqrt{(2\pi)})*\text{fnInt}(e^{\wedge}(-.5(x-7.5)^2),x,5.25,5.75)$ = .027834684209; Ans*.383 = .010660684052; Ans*50 = .533034202598] order 1 pr.

   d. [calculator screens showing: fnInt(y1,x,7.75,8.25) = .17466632194; fnInt(y2,x,2.5,3.5) = .005977036247; Ans*.17466632194 = .001043986937; Ans*50 = .052199346866] normally none, but for a best customer - 1

3. $\displaystyle\int_{x-.25}^{x+.25} \frac{1}{\sqrt{2\pi}} e^{-.5(x-10)^2} dx \;*\; \int_{w-.5}^{w+.5} \frac{1}{\sqrt{2\pi}} e^{-.5(x-0)^2} dw$, possibly no length adjustment, width D=0

[calculator screens showing: fnInt(y1,x,10.25,10.75) = .17466632194, Ans→L = .17466632194; fnInt(y2,x,-2.5,-1.5) = .060597535943, Ans→W = .060597535943; .17466632194*.060597535943 = .010584348722; Ans*300 = 3.17530461654]